North Carolina
First in Flight
1903

Team USA
Project: "Good Karma"
2018

Building Team USA

Project: "Good Karma"
USA, NATO, China, Russia, North Korea, Canada, Mexico, et al.

Team USA
A publicly presented Nuclear Disarmament Program to build an aircraft/spacecraft designed after our very own

Milky Way Universe

Author: Nick Webster

Free Agent SNW ~ Licensed

DOB: 08-06-1946, Boston

400 Money Island Drive

Atlantic Beach, North Carolina 28512

nickwebster1946@outlook.com

Ph: 970-946-3858

Briefing Phase III:

April 6th, 2018; marking our 1st Year.

NASA and our United States Armed Forces
have received Project: "Good Karma" as should be our first
objective. Phase **III** involves public participation.
This phase seeks grant development for students here in
North Carolina, Carteret Community College, other colleges,
and other universities, as with industrial participation.

To do that and more Phase **III** expands public awareness.
Public awareness and public opinion on a national
level will determine our future as

Team USA.

We can work together in the spirit of NASA and
Captain Michael Smith of Beaufort, N. C.
Captain Michael Smith was born April 30, 1945.
Captain Smith tragically perished with the entire crew of the
Challenger Mission spacecraft's departure take-off from
Kennedy Space Center on January 28th, 1986.

This is a

Great Circle Study

By

Mick

Patent Drawings Up-Date: Public Review; Project: Good Karma
USA Corporate/Government Funding Requested:
Contractor: Free Agent, Steven Nichols Webster
Drafted: March 05, 2017
Drafted Up-Date: August 06, 2018
Team USA

FIG. 5-A-1

The planet alignment below was inspired as having seen
Mars, Saturn, and Jupiter and our moon across
our evening sky on August 1st, 2018.

Earth   Mars   Saturn   Jupiter   Venus   Sun

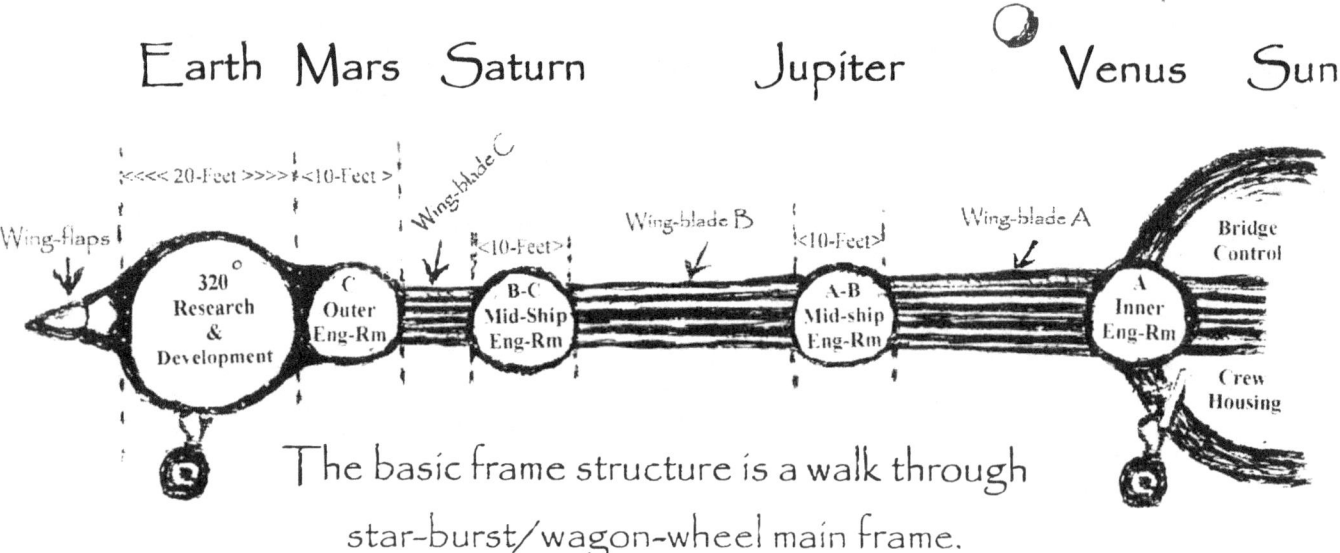

The basic frame structure is a walk through
star-burst/wagon-wheel main frame.

I could not see Venus in that evening sky. Yet, I checked
the internet and found Venus just west of our moon.
Likewise, you will work with many micro/macro ratios
derived from observations within our universe; the
Milky Way Galaxy

Where are the fish?
Where is the fish story?

First things first! You can not read a book and fish at the same time, lest the fish are not biting. Right?

Timing is everything.
Well, let's read now and fish later.

We will be covering an aeronautical flight "R&D" procedure inside Unknown Physics. This is a threshold reading opportunity for all ages, especially here in the First in Flight state of North Carolina.

Project: Good Karma is just like Eagleworks, Johnson Space Center, NASA, Houston, where Dr. Harold "Sonny" White is publicly developing an unknown physics formulary for Warp Speed. This NASA based public physics formulary class for the study of Warp Speed is the key intellectual objective giving support for the study of Project: "Good Karma"; also a study in unknown physics. Basically the macro application of the warp speed formulary is for space travel outside our atmosphere. While the micro application of such physics will work well for the here and now Project: Good Karma's self generating electric fuel sustaining flight inside our atmosphere. Now you know why all this comes down to a good days fishing on the Oceanana Pier. Everybody knows talking about the here and now makes for a good day's fishing. Yes, talking about the aircraft/spacecraft of Project: "Good Karma" is the first line of support to see this First in Flight mission gets the attention it needs. Project: "Good Karma" needs you in the big picture.

I ask our U. S. Armed Forces; Pentagon, to consider this Project: "Good Karma" as a real 20-year "R&D" nuclear disarmament program. Our civilian computer based "R&D" programs will start with a star-burst/wagon-wheel walk through frame, much like Star Wars props and other theatrical space props. This is a public presentation expected to become an active U. S. Armed Forces enlistment reality. However; an argument seems to be engaging between some Star Wars realist expressing the desire for the entire Pentagon "R&D" funding budget going to real warfare first and only. Forgive me for arguing for Peace and say: "Yes", for the public computer based Project: "Good Karma" "R&D". Think of North Carolina's Universities and North Carolina's Colleges. Then think of NATO and beyond. **All that education for the cost of one <u>Aeronautical Engineering Computer</u>.**

This is a good business plan !!!!

Variables

The variables specific to our mission aircraft/spacecraft are the space needed to house this unknown technology and the weight of that unknown technology between our wing-blades and frame. We have to design new applications for known technology in an almost opposite alignment of the known technology used in our contemporary jet engines. This state of "Does not exist yet." is not an obstacle. It is an inspiration to achieve this unknown technology with the help of our requested NASA based Aeronautical Engineering Computer. "Imposable" is not acceptable in a world that has toured our universe with known technology. The fact that this new technology is not a military advantage in its present form is not a civilian disadvantage.

Any and all micro or macro alternative applications of magnetic bearings, maglev technology, metallurgy, electro magnetic field transfer "R&D", etc. will be studied under our USA National Security interests. As an international participation is anticipated there is a normal "Top Secret" stigma binding standard national security oversight vs. a public "R&D" inside unknown technology "R&D". Working from a NASA based Aeronautical Engineering Computer is not the same as Wilber and Orville Write becoming the first in flight in 1903. The difference is having over 100 years of advancing flight "R&D" on our side in our USA.

PEACE is our technological objective while filling the vacuum caused when other countries cease their nuclear warhead "R&D". Farming incoming meteorites is our extended international application for this new technology and new mission aircraft/spacecraft.

Say "Yes" and be a voice for Team USA.

Phase III "R&D"
Requested: NASA's Flight Probability Analysis Objectives

1st: Develop a computer based Flight Probability Analysis program capable of serving public education; local input, with any and all concentric formations of engine rooms, power systems, research areas, flight control, passenger seating, and whatever else students ask our star-burst/wagon-wheel frame to house and fly; submitted in Phase I. This will be easy to adapt ideas at the designer end without material expense.   Engine Rooms/Observation Areas/Frame

2nd: Develop that computer based program to assist in designing the drive-beams submitted in Phase I. The drive-beam function is to ride within the maglev process and thereby lift the aircraft/spacecraft into sustained flight. The round/tubular drive-beam is technologically more difficult to manufacture/produce than a flat against flat maglev contact area. Thus both drive-beam contact application studies will be available.

3rd: Develop that computer based program to assist in designing the maglev bearing process as submitted in Phase I; including both 6-inch to 10-inch flat contact areas and the rounded maglev contact areas.

4th: Standardize all materials specific to the moving parts and contact areas within a contemporary jet engine towards like functions within our mission aircraft/spacecraft. Add an option window to appear when the use of a "to be developed" alloy, carbonate, etc. whenever appropriate.

5th: Request our U. S. Congress to OK the laser based throttle study.

Entering Unknown Physics:

Basically I started designing a flying generator powered with a helicopter engine. I then knowingly burdened that helicopter engine with much more than it was originally intended to do. This start-up has no public test record. I hope and trust my first patents went to Area 51 for "R&D".

Today my hopes and dreams are with our United States Congress and private investment. President Trump has set a 2020 window for the formation of a United States Space Force. Our Pentagon will decide what they need and what they want. Will the Pentagon want a non-nuclear powered aircraft/spacecraft to represent our International Nuclear Warhead Disarmament proceedings? I certainly do. Yes, the maglev tests and the actual wing-blade flight capability tests will start with contemporary fuel and contemporary power systems. As we generate enough electricity to power both our maglev bearings and electronic throttle we have achieved a milestone in flight. Our long term goal to become self powering and self sufficient; equilibrium of mechanical motion sustaining flight is nearing completion. There and then we feed an electronic laser main-drive propulsion to the drive-beam itself; an unknown process. Then we program a desired 30,000 RPMs {+ or -} for VTO and sustained flight.

Advanced Computer Flight Probability Studies

This should be a Spacex, NASA, Boeing, et al Think Tank Project.

Helicopter Start-Up

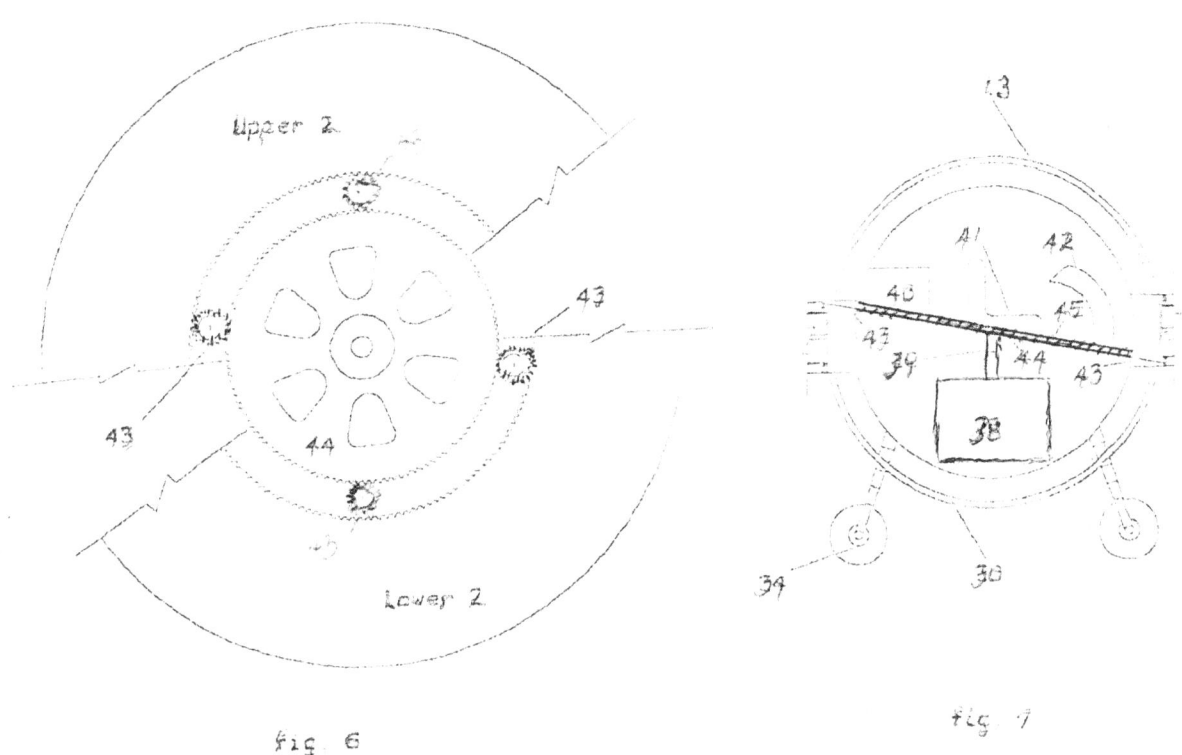

Upper 2

43

43

44

Lower 2

fig. 6

43

41   42

40

45

43

39   44   43

38

34   38

34   30

fig. 7

At first the pilot sat on top of the 1st stage fly-wheel. Then a flight team sat in the outermost flight research area. This start-up power system has up-graded to contemporary fan jets housed with swivel Harrier type relocation assemblies.

Now remember! The most economical start-up today is setting up an Aeronautical Engineering Computer in a North Carolina University or college dedicated to finding that most aeronautical design for a multi-concentric wing-blade unto the greater of 12 major frame extensions for either flight control, engine rooms, R&D, passengers, or storage are tested. Then we make this start-up a downloadable public education program.

#1 - Test: Test many wing-blade alignments. Then advance into the electric throttle or laser based throttle of our future years. A greater compression discharge presser will improve flight ability. Becoming weather resistant would include an upper rotating shield of sorts well above the air intake wing-blades. I am not the science guy or an aeronautical engineer. I put this aircraft/spacecraft patent to bed some twenty years ago when the Air Force said it would take a third to a half of our "R&D" funding to get an official aircraft/spacecraft in steady flight. I submitted this mission aircraft as unsolicited in a 1989 USA/USSR Nuclear Disarmament Proposal, Counter Atomic Attack System One; Operation: Cultivation of the Stars. This 2017 - 2018 I have focused on China and North Korean Nuclear Warhead Disarmament.

The Test: The two wing-blades in opposition under the engine room compartment is but one test. Several wing-blades rotating in the same direction with the outermost wing-blade balancing flight equilibrium in the opposite direction is much the same as a helicopter with a tail rotor. There will be a most successful flight formulary for this concentric wing-blade or mobile-wing aircraft/spacecraft. We must test to find it.

#2 - Parts: The most difficult part to develop is the Drive Beam. The Drive Beam is the RPM rail for the wing-blades. Those Drive Beams have multiple bearing assemblies; air, metal, maglev. Those Drive Beams support the total aircraft weight divisible only by the pairs of wing-blades we employ. The 9-Planet overview offers 7 sets of wing-blades with 8 engine-room areas. With positive computer flight probability analysis tests we can move forwards within the original international participation and cost reduction objectives.

Briefing:

On April 06, 2017 Senator Bill Nelson
forwarded Project: "Good Karma" to NASA
for a computer flight probability analysis.

August 14, 2018
That presentation is now a book named
Project: "Good Karma"
sold at Barns & Noble and Amazon

The following presentation has an extended
Great Circle Study for kids of all ages and additional
drawings of the aircraft/spacecraft sought to represent
Nuclear Disarmament all over our Earth.

President Trump has started great interest in forming a
6th branch of our armed forces named the
United States Space Force.

It is up to you! Do you want the herein disclosed
aircraft/spacecraft to be Researched and Developed ?

Project: "Good Karma"

An International Technology Sharing
"Research & Development" Project
to build an aircraft designed after our own

<u>Milky Way Universe</u>

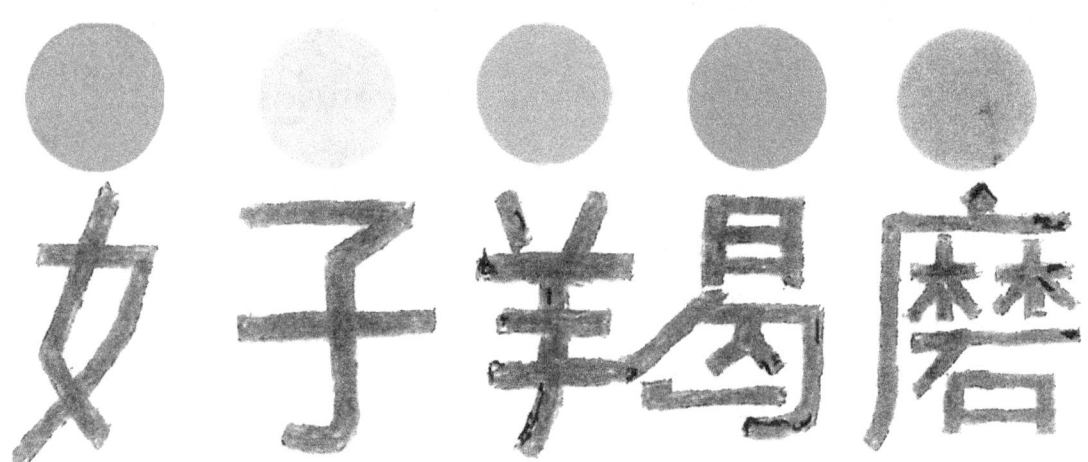

<u>Team USA</u>

Hăo jiémó

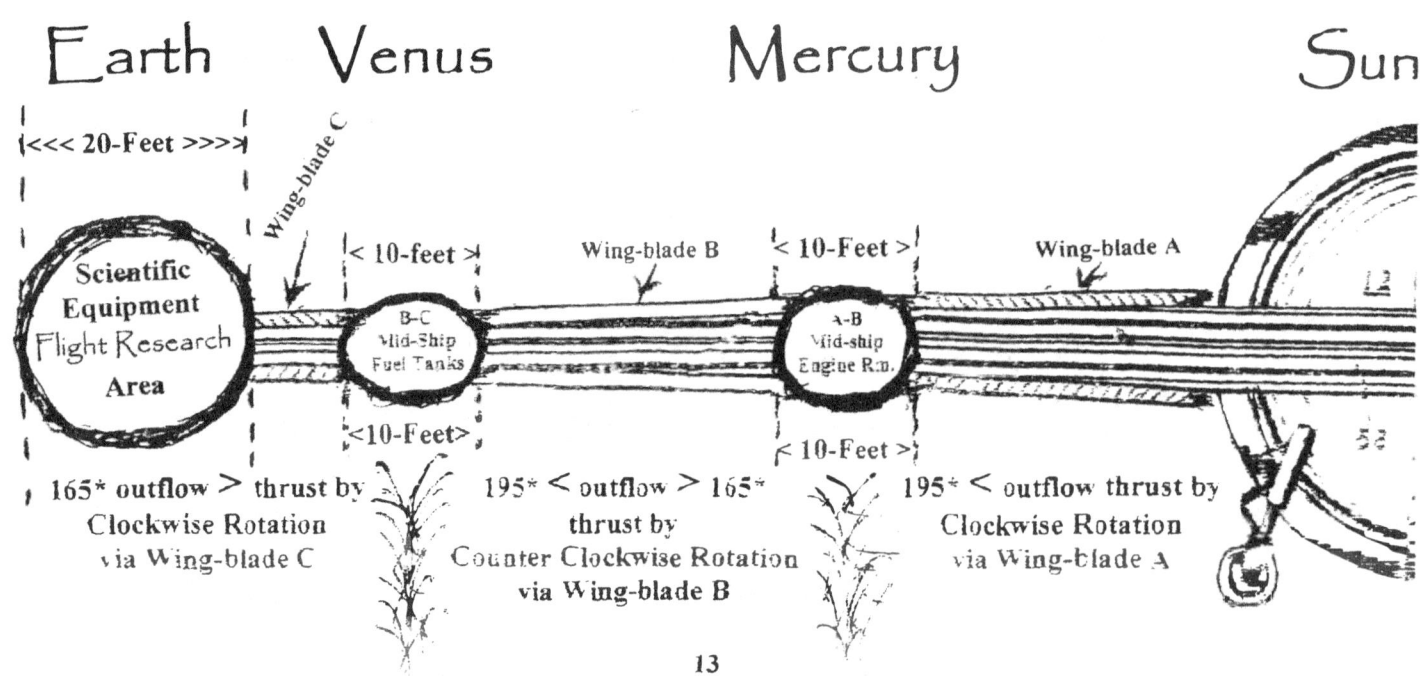

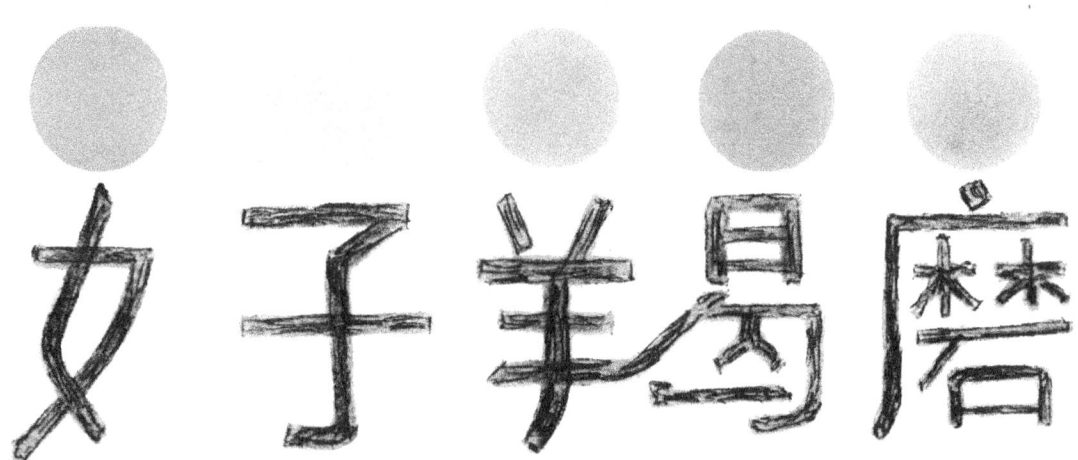

The following is an up-graded presentation of

Project: "Good Karma".

Only a few pages are original April 6th, 2017 pages.
Think about it. Improving the following
diagrams into an actual working
aircraft/spacecraft
is the future
of our

Team USA.

Prepared For:

总统　President Donald Trump

And

总统　President Xi Jinping

Presented for tabling during the
Meeting Of Presidents; April 6th, 2017

Prepared by: SNWebster

Project: "Good Karma"

This is the drawing used back on
April 6th, 2017
when speaking up for an
aircraft/spacecraft designed after
our universe.

Team USA

FIG. 5- A

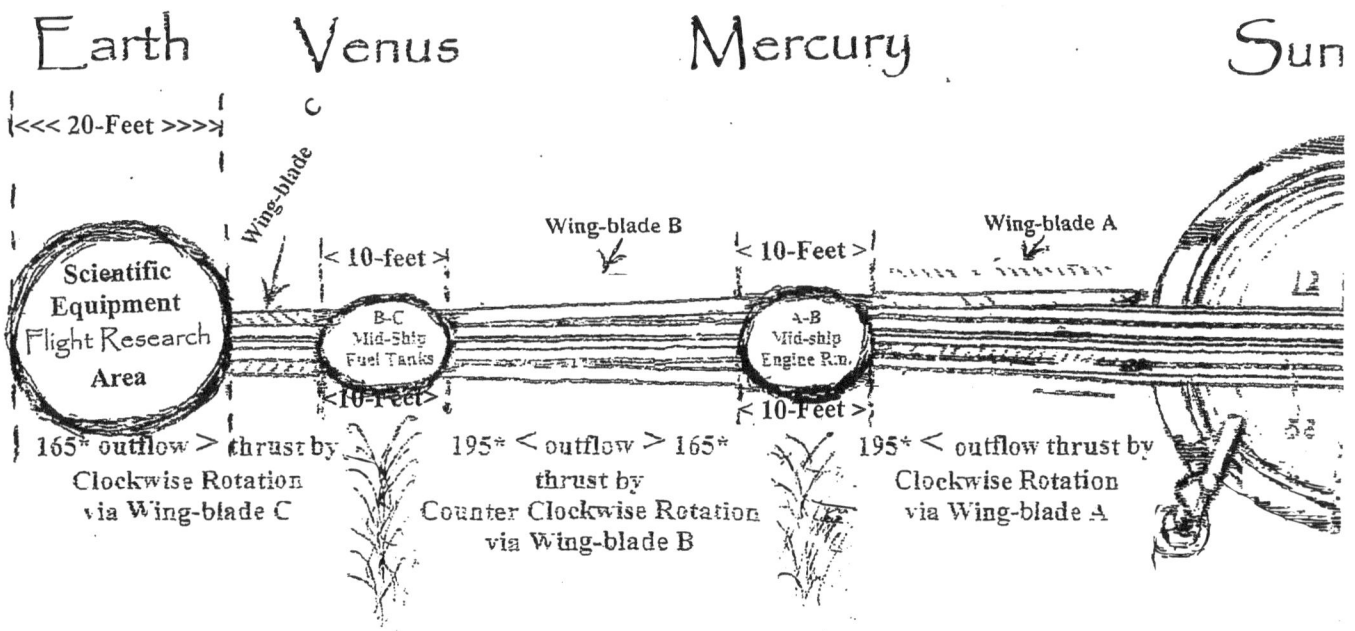

Earth   Venus   Mercury   Sun

|<<< 20-Feet >>>|

Wing-blade C

< 10-feet >

Wing-blade B

< 10-Feet >

Wing-blade A

Scientific
Equipment
Flight Research
Area

B-C
Mid-Ship
Fuel Tanks

A-B
Mid-ship
Engine Rm.

<10-Feet>

< 10-Feet >

165* outflow > thrust by
Clockwise Rotation
via Wing-blade C

195* < outflow > 165*
thrust by
Counter Clockwise Rotation
via Wing-blade B

195* < outflow thrust by
Clockwise Rotation
via Wing-blade A

Yielding a Collateral B-C
Lifting Turbulence Stabilization
or VTOL

Yielding a Collateral A-B
Lifting Turbulence Stabilization
or VTOL

The above references to 165-Degrees & 195 Degrees
are by Solar North as if the Sun at high noon were North.

Up-dated March 5th, 2017 offering of now outdated patents
U.S. Patent 5,213,284; 5/25/93 & Design Patent 320,378; 10/01/91
by S. N. Webster, DOB 08/06/1946, Boston

To the:

Honorable United States President Donald Trump
and
Honorable Chinese President Xi Jinping

Sirs, I thank my Honorable Senator Bill Nelson for bringing this Project: "Good Karma" before you at this time. I wish to inspire children and adults alike to dream, study, and overcome "Eternal War". The aircraft documentation introduced here-in seemingly separates my humble beginnings with paper and pen from completion with the term { **Expired** }. Project: "Good Karma" can be completed without my oversight. I simply offer my support for the completion of Project: "Good Karma" at this time via this meeting of our United States President Trump and Chinese President Jinping.

To my corresponding Chinese team,

Your maglev technology is big thunder. I do not yet understand everything you have done. Congratulations! I ask you. Can you make this now out-dated by patent coverage; VAu 48-797, mobile wing tricentric displacement aircraft fly with your maglev technology? I can only imagine how. Our countries and peoples should technologically evolve together.

Respectfully yours in Christ @ Sea & @ Home.
With Peace of Mind.

Steven Nichols Webster
Steven Nichols Webster

Project: "Good Karma"

April 03, 2017

Purpose: USA/China Technology Sharing

Program: Research & Development, Aviation

Objective: Build an aircraft designed after our very own
           universe the Milky Way by width and depth.

Reason: To build a "Vehicle of Peace" in thought, word, and
           deed; ownership investment by participating nations.
           "All Nations Invited."

History: November 15, 1989
           I; S. N. Webster, authored a USA/USSR
           Nuclear Disarmament Proposal Counter Atomic
           Attack System One; Operation: Cultivation of the
           Stars. There-in I offered the original patents that this
           Project: "Good Karma" is centered on as the mission
           aircraft. CAASO; Op: CS was a predecessor to
           the International Space Station we know today.
           Today, Chinese maglev bearings may just be the
           technology needed to get Project: "Good Karma" off
           the ground.

                                          by   S. N. Webster

Here we have the **Tri-Centric** arrangement of wing-blades. This was my first entry. Wing-blade "B" has a snowplow blade sending intake in both outer directions. That way equal amounts of intake will produce equal forces in opposition below.

However, the chance of that tri-centric configuration being the most useful concentric figuration in the world of physics and flight is very low. Yes, that would be about like being the only civilized; evolutionary included, planet in the galaxies beyond our Milky Way galaxy.
There are many concentric configurations to test.

By area and weight
A + C = B
A = C

By area only
A + B + C = 80% r6

By width only
d = e
r6 + 2d = r7
r6 + 2e = r7

That is why an

Aeronautical
Engineering
Computer

is a good business plan.

Team USA

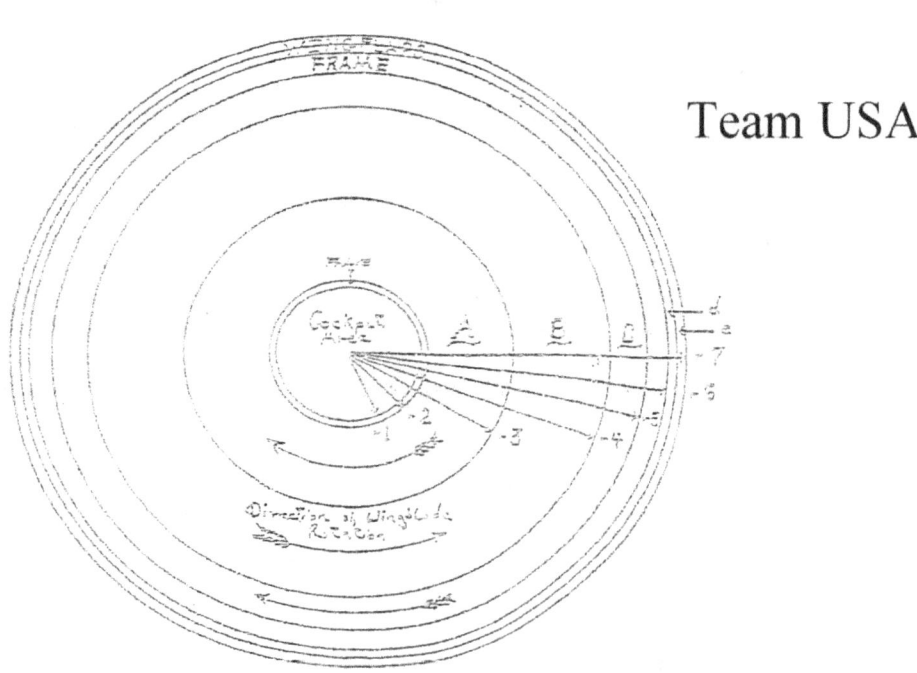

FIG. 3 - B

Patent Drawings Up-Date: Public Review; Project: Good Karma
USA Corporate/Government Funding Requested:
Contractor: Free Agent: Steven Nichols Webster
Drafted: March 05, 2017
Heading Up-Dated: August 06, 2018

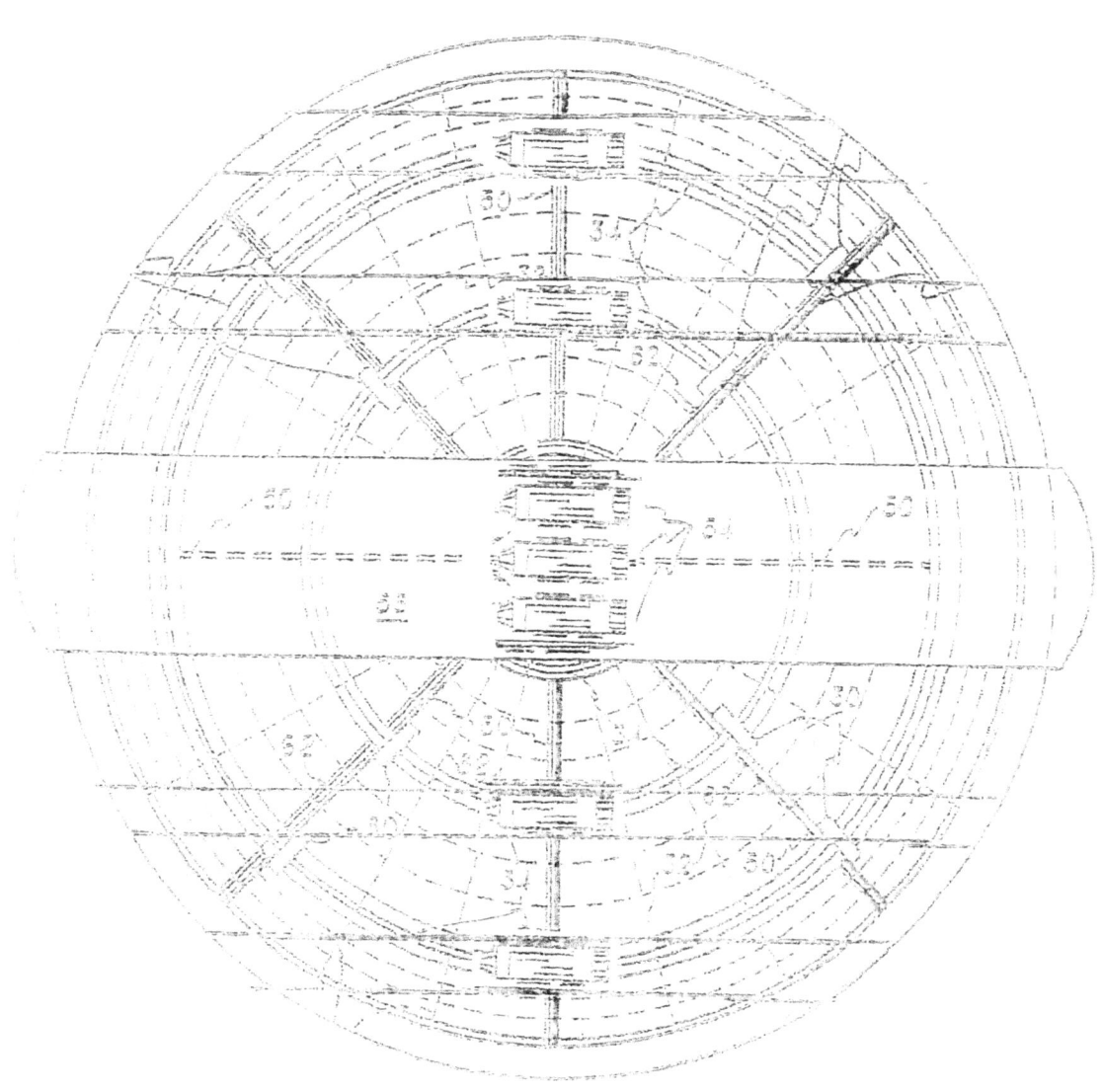

*Turbo Jet Thrusters are affixed
to the BC Frame and to the CB Frame*

FIG. 4-A

Patent Drawings Up-Date: Public Review; Project: Good Karma
USA Corporate/Government Funding Requested:
Contractor: Free Agent: Steven Nichols Webster
Drafted: March 05, 2017
Up-Dated: August 06, 2018
Team USA

FIG. 5-A-1

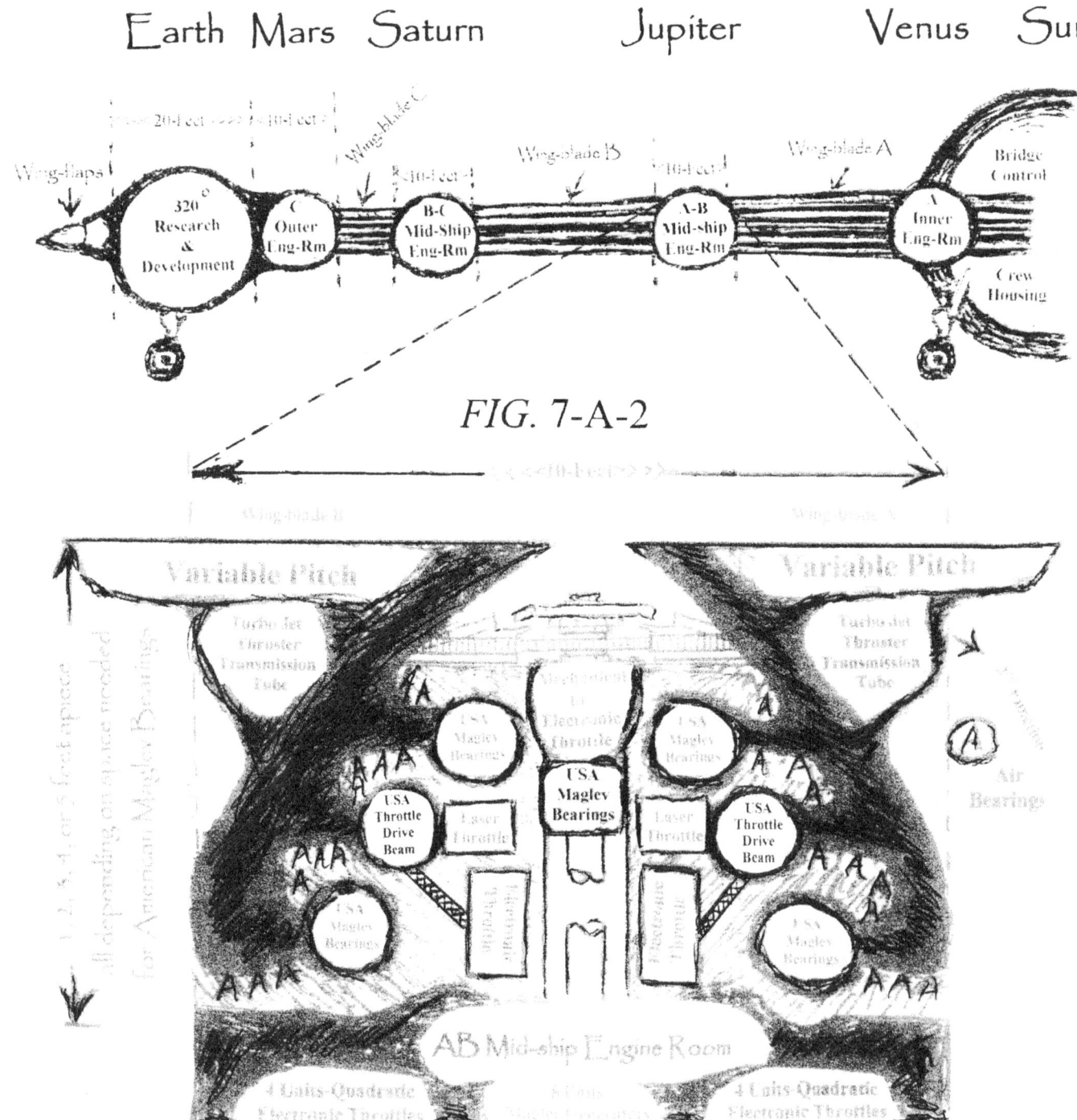

Earth   Mars   Saturn   Jupiter   Venus   Sun

FIG. 7-A-2

Patent Drawings Up-Date: Public Review; Project: Good Karma
USA Corporate/Government Funding Requested:
Contractor: Free Agent: Steven Nichols Webster
Drafted: March 05, 2017
Up-Dated: August 06, 2018
Team USA

FIG. 5-A-1

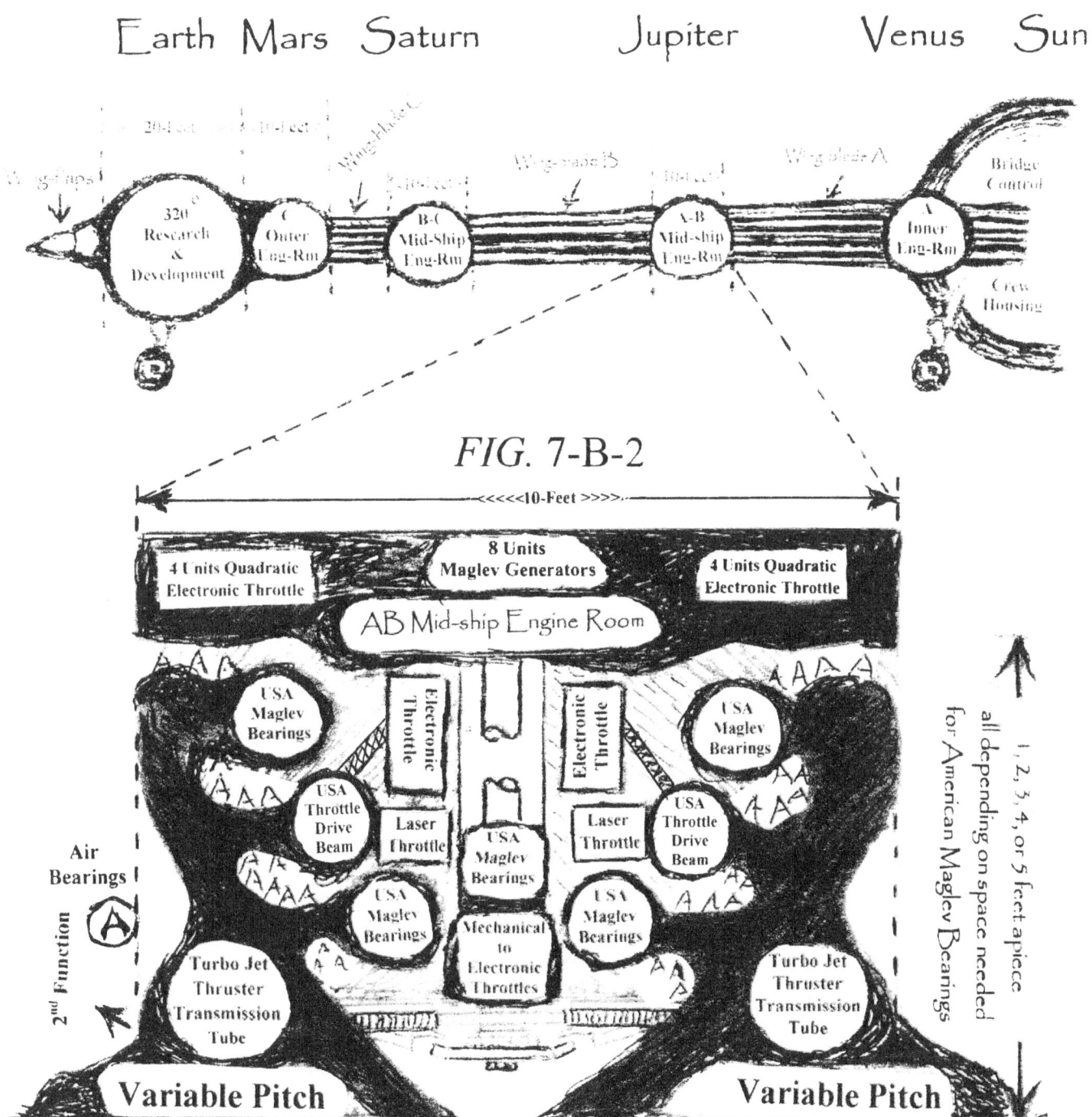

Earth Mars Saturn   Jupiter   Venus Sun

FIG. 7-B-2

Patent Drawings Up-Date: Public Review; Project: Good Karma
USA Corporate/Government Funding Requested:
Contractor: Free Agent: Steven Nichols Webster
Drafted: March 05, 2017
Up-Dated: August 06, 2018
Team USA

FIG. 5-A-1

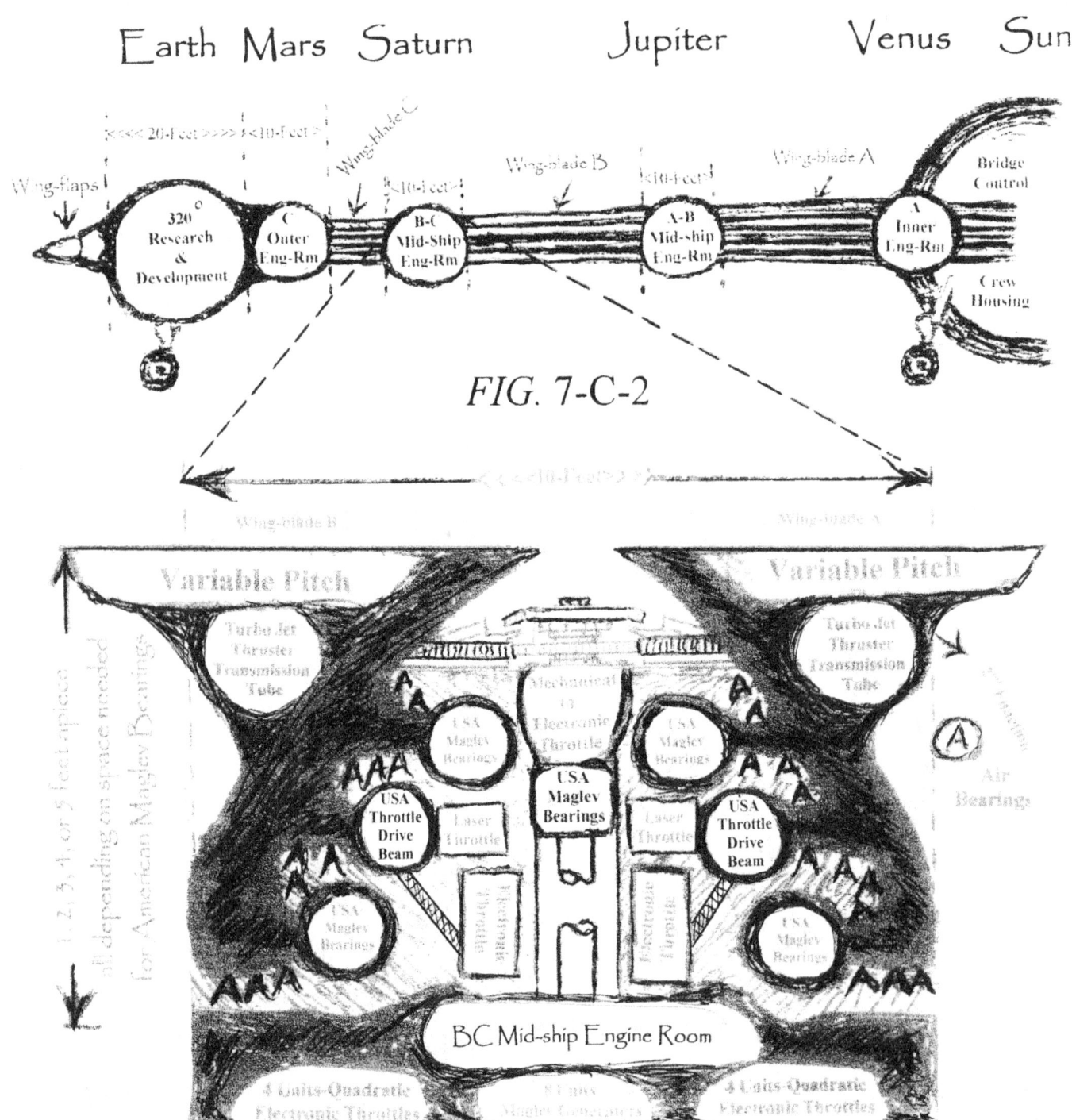

Earth Mars Saturn   Jupiter   Venus   Sun

FIG. 7-C-2

Patent Drawings Up-Date: Public Review: Project: Good Karma
USA Corporate/Government Funding Requested:
Contractor: Free Agent: Steven Nichols Webster
Drafted: March 05, 2017
Up-Dated: August 06, 2018
Team USA

FIG. 5-A-1

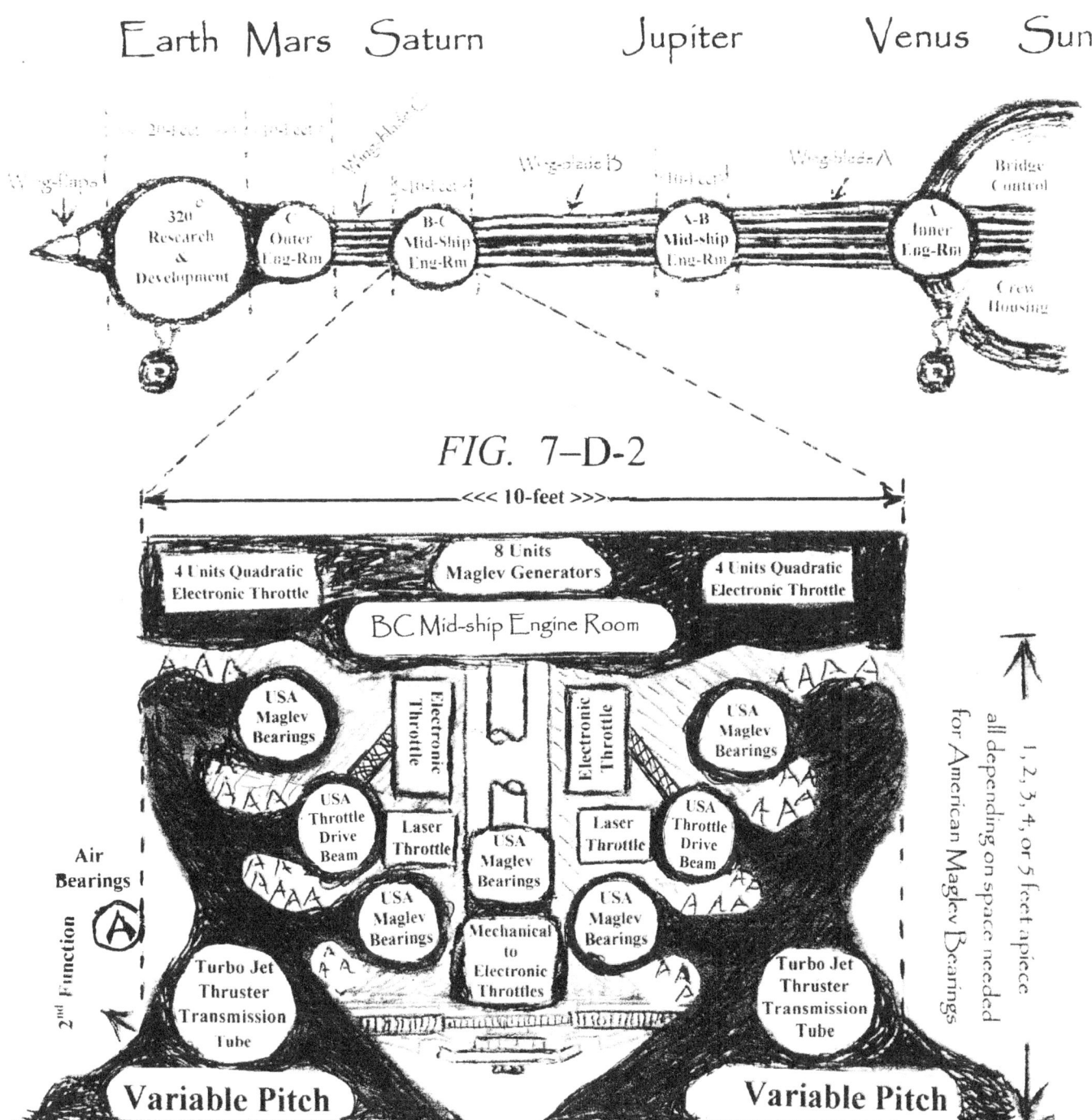

FIG. 7–D-2

Briefing Phase II:

April 06, 2017

Logically, I seek Pentagon interested people as the money required to develop this still unproven flight technology; Project: "Good Karma", will go to both NASA and our United States Armed Forces as we step forwards.

August 14, 2018

This presentation is a short version of
Project: "Good Karma"
written for our local fishing community and kids everywhere. I love to take a cup of coffee out on the Oceanana Pier and sit just above the shore break when I stay in Atlantic Beach. I find listening to the shore break on the Oceanana Pier in Atlantic Beach, N. C. very relaxing.

Fig. 5-A-9 is a Mother ship without the star wars theme. Fig. 5-A-9 is an earth bound aircraft/spacecraft designed after our own universe the Milky Way. Again the most difficult concept to put together are the Drive Beams that ride within the maglev bearings. They are the upper and lower of the three. The middle drive-beam is the throttle drive-beam. This centermost drive-beam must carry enough copper to transmit electricity. The copper is a timing regulator for the laser throttle. The outer two; upper and lower, must be magnetic to balance within the maglev bearings.

Evolution of the Drive Beam Throttle Concept: I first thought of a yard stick made of light iron ¼" x 1" x 36" twisted once by 360-degrees, then reconnected end to end. Then we fill the void with copper. Then I thought of twisting that same drive-beam several times by 360-degrees then reconnecting. Once done we round the circular/outer face with a strong sheet of contact alloy facing the metal and air bearings. Next "R&D" is designing a balanced number of outer pockets into which less than weapon grade lasers can be used for propulsion/fuel to reach 30,000 RPMs. Now start with a processed 5 point star drive beam that is also twisted 360-degrees; once or several times, and reattach. Again fill the voids with copper, add a strong magnetic outer face, and finish with pockets suited to absorb less than weapon grade laser propulsion. Think of animation supplementing this design process of a multi-alloy multi function drive-beam. Animation, artificial intelligence, and the best computer programmers NASA, Pentagon, and Free World have to offer will fly our mission aircraft/spacecraft for millenniums to come.

Overcome "Eternal War"
with a work of technological art designed after
our very own Milky Way Universe.

Patent Drawings Up-Date: Public; Project: Good Karma
USA Corporate/Government Funding Requested:
Contractor: Free Agent: Steven Nichols Webster
Drafted: March 05, 2017
Up-Dated: August 06, 2018
Team USA

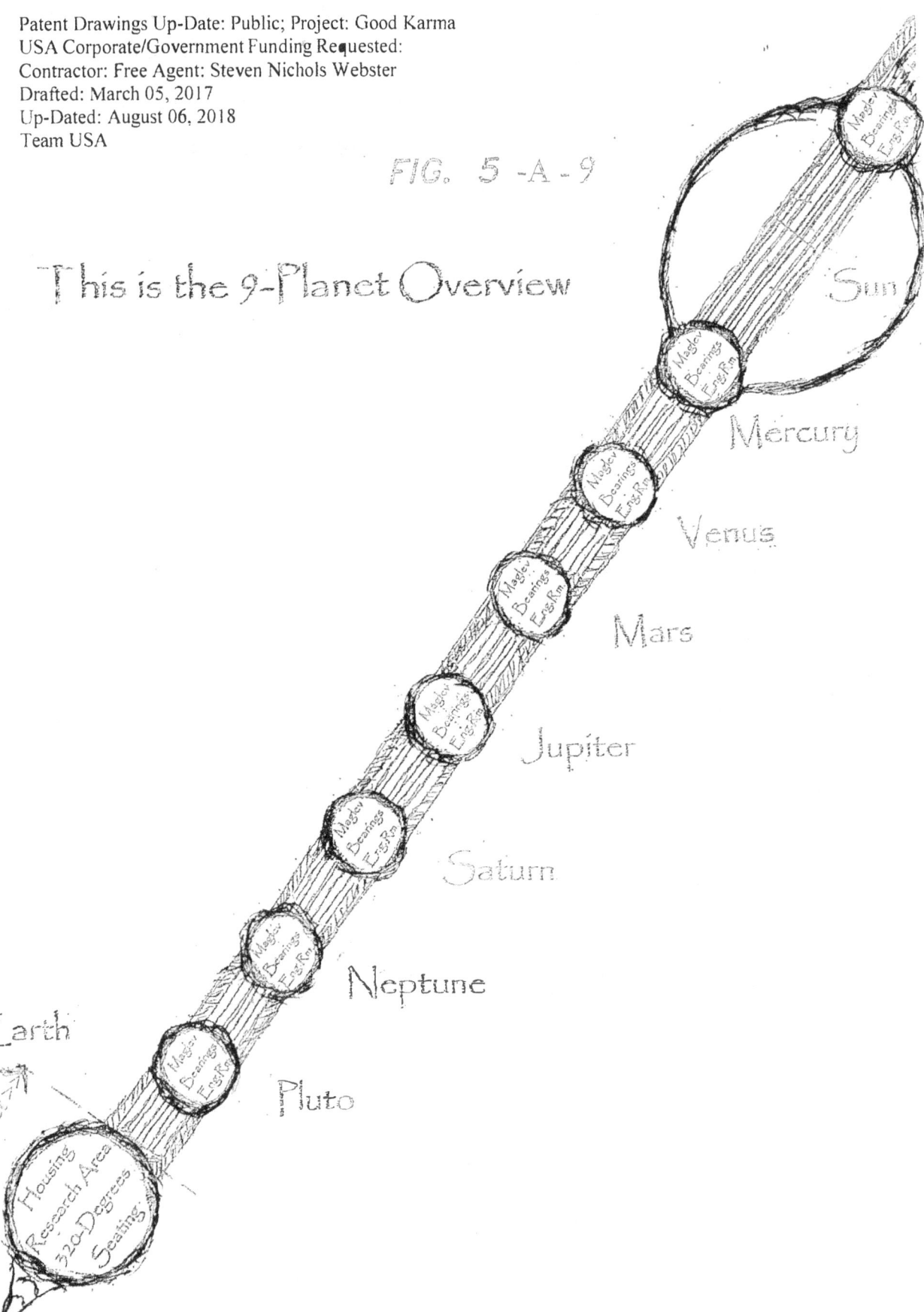

FIG. 5 -A - 9

This is the 9-Planet Overview

Senator Bill Nelson
413 Clematis Street, Suit #210
West Palm Beach, Florida 33401
Ph: 561-514-0189
Fax: 561-514-4078

April 03, 2017

Steven Nichols Webster
800 Uno Lago Drive, #203
Juno Beach, Florida 33406
Ph: 970-946-3858
DOB: 8/6/1946, Boston

Reference: USA/Chinese Technology Sharing.
Project: "Good Karma", Tricentric Aircraft VAu 48-797.
Do share Project: "Good Karma" with President Donald Trump
before Chinese President Xi Jinping arrives here in West Palm Beach.

Dear Senator Bill Nelson,

Sir, I thank you for your life of successful NASA missions and political
representation here at home. I offer this Project: "Good Karma" towards
China/USA relations while seeking Russia/NATO/ Etc.
participation. Please help me bring this Project: "Good Karma";
mechanical to electrical flight off the ground while Chinese President Xi
Jinping comes to West Palm Beach to visit with President Donald
Trump at Mar-a-Lago. Please represent me; a constituent. Our
USA/CHINA relations need constant care. This is more than just
one more project. You know my interest in NASA's Cassini May 2nd &
3rd, 2013 observations of a time I pray that never encompasses Earth
with the same outcome. Please table Project: "Good Karma" with
President Donald Trump before President Xi Jinping arrives.

Respectfully yours in Christ @ Sea & @ Home.
With Peace of Mind.

Nick Webster

Fax To: **202-636-0711**   April 06, 2017

FedEx Washington, D.C. Management   Nick Webster

snwebster@att.net

970-946-3858

Ladies and Gentlemen,

Long story short, Senator Bill Nelson sent Project: "Good Karma" to NASA not to the President Xi Jinping and President Trump meeting. Therefore, my earlier request to you folks to resend my PACKAGE Tracking # 786148945377 to the Chinese Embassy in D.C. would be up-staging Senator Nelson's NASA decision. I do NOT want to up-stage Senator Bill Nelson. Please, do not resend #786148945377 to the Washington D.C. Chinese Embassy. Thank you.

I will pay the difference tonight here in Juno Beach to re-lable that package in your D.C. facility to the following address:

Mr. Jared and Mrs. Ivanka Kushner
"American Innovation"
The White House
1600 Pennsylvania Ave. N.W.
Washington, D.C. 20500

They will understand and I will thus follow Senator Nelson's lead. The Chinese would prefer they receive their invitation to Project: "Good Karma" in Chinese. That would take me years.

Respectfully yours in Christ @ Sea & @ Home.
With Peace of Mind.

Nick Webster

NASA Headquarters   March 7th, 2015
Suite 5R30   S. N. Webster
Washington, D.C.   351 Zenith Lane
20546   Juno Beach, Fl. 33408
Fax: 202-358-4338   561-635-2847

Dear NASA,

Kind Sirs, I have over extended myself in an observation of my own before checking with you of NASA to verify the observation is NASA verified. I used my interpretation of a May 2nd and May 3rd , 2013, Cassini, filming of the super storm on the North Pole of Saturn. My observations were before the hexagram observation that seems to be the apple of the moment over the internet. Allotropic configuration of minerals displaced by the storm I presume. Therein, I have not been able to relocate the filming in question that I did observe via the internet at least 10 times over the years.

Now, I am responsible for my words. I am responsible because I wrote about this observation as background material in an ally building document I sent to Speaker of the House John Boehner just last week. A hard copy will be arriving by mail; as per any internal request for verification.

Therein, I said this: "We need a ***Common Denominator*** in our dialogue with Russia. **Money is a good common denominator.** Our USA Industrial Complex has matured into space travel trajectory and accomplishments beyond expectation. Just for this moment keep your thoughts "Out of this world". Look at the May 2nd and May 3rd , 2013 NASA Cassini coverage of the super-storm on the North Pole of the planet Saturn., That hurricane spun or now spins in two directions at once. Truly, it is out of this world. That combined counter-clockwise storm coupled with a clockwise storm of equal center maximum force seemed to gather enough force to reconstruct every molecule of what was once on solid ground. Perhaps water as we know it today had been on planet Saturn some 20 times longer than water has been here on earth and perhaps people there never learned to get along. This is my closing though on mending bridges between Russia, the USA, and the Middle East overcoming Eternal War."

#1: Before the observation of the magnetic hexagram or hexagram discussion of Cassini May 2nd and May 3rd , 2013 was there NASA

recognition of a super-storm on the North Pole of planet Saturn that did earnestly spin in two opposite directions at the same Time.
{{{{ Yes or No }}}}

#2: I never mentioned this #2 issue as I did wonder if NASA under first review did observe red the same. My observations went ever farther in seeing that same "Out of this world" super-storm. I actually saw an energy source above the storm that looked like a living donut made of winds and light that spun from the outside to the inside around and around at a speed that was incredibly beyond, seemingly wind approaching the speed of light. It hovered above the storm like a jelly fish only compact yet transparent and wobbling within the tandem motion between it and the super-storm. The donut shaped energy was much smaller than the storm. The energy wobbled as the spinning occurred around the shorter diameter of the donut moving in an up the outside and down the inside. The winds moved from the outside to the inside around the intercept rather than around the donut 360 shape. The 360 shape would be as it laid on a table, so to speak.

I know you have seen God in action as you have observed time as you have. I am honored to live in a time as NASA has honored we the population of earth with such incredible achievements as transferring observations from so far away. Evolution has occurred.

Change of Subject: Many good friends have asked me: "Where is the money going to come from?" I am speaking of money to employ beyond the standards of the Gold Standard. NASA, you are showing the world where the money is going to come from. I was writing Mr. Speaker John Boehner about that money to employ a growing world as we overcome Eternal War.

Respectfully yours in Christ @ Sea & @ Home.

With Peace of Mind.

Steven Nichols Webster

Chairman Senator Richard Burr   March 4th, 2018
U. S. Senate Select Committee on Intelligence
211 Hart Senate Office Building   Steven Nichols Webster
Washington, D. C.   20510   400 Money Island Drive
202-224-1700   Atlantic Beach, N. C. 28512
                            970-946-3858

Reference: Project: "Good Karma" requests your support Senator Burr.
North Carolina is and always will be the "First in Flight". Please help me
get this mission to manifest off the ground in North Carolina style; in the
public eye.

The Honorable Senator Richard Burr,

Sir, I was schooled by the Outer Banks' very own Captain Jim Zook of
Morehead City in 1984. That same November I tested and received my first
USCG Captain's License. My medical discharge from the Military Sealift
Command in 2009 is included in the attached Project: "Good Karma". I am
Nick Webster, I ask for your support Sir. Many truths are self evident. This
time sensitive international objective of "Overcoming Eternal War" is a
reality. President Donald Trump has yet to comment on my project.

Change of subject: Did you hear that this March 2018 marked another
first; the first temperatures above freezing were recorded on our North Pole.
In this waning of this Ice Age civilization has produced all written languages
known today; be they written in stone, or clay, or on papyrus, and paper,
even now on smart phones. Sir, I want my Project: "Good Karma" to
become a household conversation. China and Russia would love to take part
in this project. Switzerland has been working to produce "anti-mater' for
almost a decade. Their dreams becoming reality would make my dreams all
the easier to get off the ground. In that light I can only hope our USA Team
in Area 51 has flown an adaptation of my tricentric U.S. Patent # 5,213-284
as per my U.S. Design Patent #320-378; as having already built an
aircraft/spacecraft designed after our very own spiral galaxy the Milky Way.

Respectfully yours in Christ @ Sea & @ Home.

With Peace of mind,

Nick Webster   *Nick Webster*

Dedicated to my brother

Captain Kirwin Shedd Webster

Capt. K. Shedd Webster, USN; during the Viet Nam War.
NROTC 1962-67, Commissioned 1967, Retired 1993.
Flew A-4 and A-7 aircraft primarily. 3000 + Hours,
600 + carrier landings and 100 + combat missions.

My brother Shedd is alive and well with his wife Pam in Colorado.
Shedd now works for a Ski Resort and lives for his kids and grandkids.

Balance

With balance in our hearts, minds, souls, and nations the here-in Project: "Good Karma" begins.

The End

Briefing: Recent Public Discussion {1-on-1} on Project: "Good Karma"

Two very respectable friends of mine brought out two very direct and seemingly popular opinions while discussing the logic of my success in my international effort to build an aircraft/spacecraft designed in the likeness of our very own spiral galaxy the Milky Way. Why symbolize our nation's efforts in "Overcoming Eternal War" technologically? We will discuss funding concepts after these other two points of view have been established as #1: and #2:. Funding builds the aircraft/spacecraft.

#1: 2018 Statement against "Overcoming Eternal War".
#1: <> "There are simply too many people on Earth. The population of Earth doubled just a few decades ago." My rebuttal: "I do want to employ those generations to come; those now considered an oncoming over-population problem. It will take more than the Gold System was built for to employ a maturing population.

#2: 2018 Statement against the logic of "Overcoming Eternal War".
#2: <> Christ said: "There will always be wars and rumors of wars."
My rebuttal: "I am a Christian by choice and by life experience. Christ also spoke of the End of the Age and said "You will always hear about wars and rumors about wars... Be not alarmed." Every language known to mankind be it written in stone, or on papyrus, or parchment, paper, or on i-phones was and now is written in the waning of the Ice Age. On the North Pole this February and March of 2018 were the first temperatures above freezing ever recorded. I know this is a change of subject. However, it will take a lot more money rebuilding both the private and public sectors than we have planned for if Mother Nature rains, snows, and blows all our polar zones back our way. I prefer to pray for a gentle transition to frozen equatorial atmospheric rings like Saturn.

The End and Some More

You have 6 more pages of original start-up patents plus
1-futuristic drawing named "Endless" before we change subjects.

Next, we review 5 letter copies that tie together my
September 20-26, 2018; Washington, D. C. presentation of
<u>Building Team USA</u>.

Now we change subjects from our Aircraft/Spacecraft
to the beginning of our
<u>Great Circle Study</u>
with mechanical time, compass, and navigation
for our younger generations. Grades: K – 6 – 12 – Captain.

<u>You millenniums take notice !</u>

Your next Aeronautical Engineering Computer start-up can become
an advanced system with 7-vertical sets of internal reduction blades
compressing the uppermost intake and ramping up that discharge
pressure like the inside of a conventional fan jet. Also try a flight
probability analysis with a single wing-blade instead of my original
2 wing-blades; one top and one bottom. A one wing-blade system
will be easier to build a computed flight probability study there-of.
Then we can multiply the sets of wing-blades per engine room
as we did when we were developing the original jet engine.
There will be a best way to test and to fly our mission
Aircraft/Spacecraft.

<u>Team USA</u>

United States Patent [19]

Webster

[11] Patent Number: 5,213,284

[45] Date of Patent: May 25, 1993

[54] **DISC PLANFORM AIRCRAFT HAVING VERTICAL FLIGHT CAPABILITY**

[76] Inventor: **Steven N. Webster,** P.O. Box 426 Sleepy Hollow, Long Creek, Mossy Head, Fla. 32434

[21] Appl. No.: **772,904**

[22] Filed: **Aug. 5, 1991**

Related U.S. Application Data

[63] Continuation-in-part of Ser. No. 395,358, Aug. 17, 1989, abandoned.

[51] Int. Cl.⁵ B64C 29/00
[52] U.S. Cl. .... 244/23 C; 244/12.2
[58] Field of Search .... 244/23 C, 12.2, 23 B, 244/53 R, 60, 17.19

[56] **References Cited**

U.S. PATENT DOCUMENTS

2,863,621	12/1958	Davis	244/23 C
2,935,275	5/1960	Grayson	244/23 C
3,312,425	4/1967	Lennon et al.	244/12.2
3,774,865	11/1973	Pinto	244/23 C
4,014,483	3/1977	MacNeill	244/23 C
5,039,031	8/1991	Valverde	244/12.2

FOREIGN PATENT DOCUMENTS

0678700	1/1964	Canada	244/23 C
2648504	2/1978	Fed. Rep. of Germany	244/23 C

Primary Examiner—Joseph F. Peters, Jr.
Assistant Examiner—Christopher P. Ellis
Attorney, Agent, or Firm—Richard C. Litman

[57] **ABSTRACT**

An aircraft having a generally circular or disc planform configuration provides the capability of vertical flight through two concentric sets of lifting fans or blades. The two sets may each include a number of individual rings of blades, but both sets are equal in area and rotate oppositely in order to provide nearly equal volumes of airflow, and thus essentially offset any torque reaction due to the rotation of the blade sets. Several engines are provided in the preferred embodiment, with one engine providing power to the lift fan sets and other engines providing thrust for horizontal flight. Other novel features are also disclosed, such as a peripheral aerodynamic control system, power transmission system, and surface vane system. An alternate embodiment includes a peripheral passenger or cargo area, with more conventional rearwardly located aerodynamic controls for horizontal flight.

15 Claims, 5 Drawing Sheets

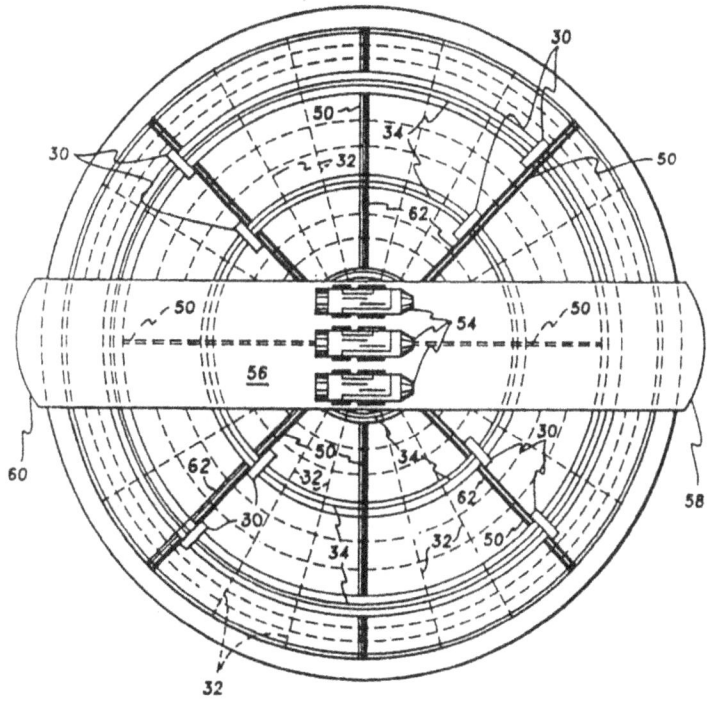

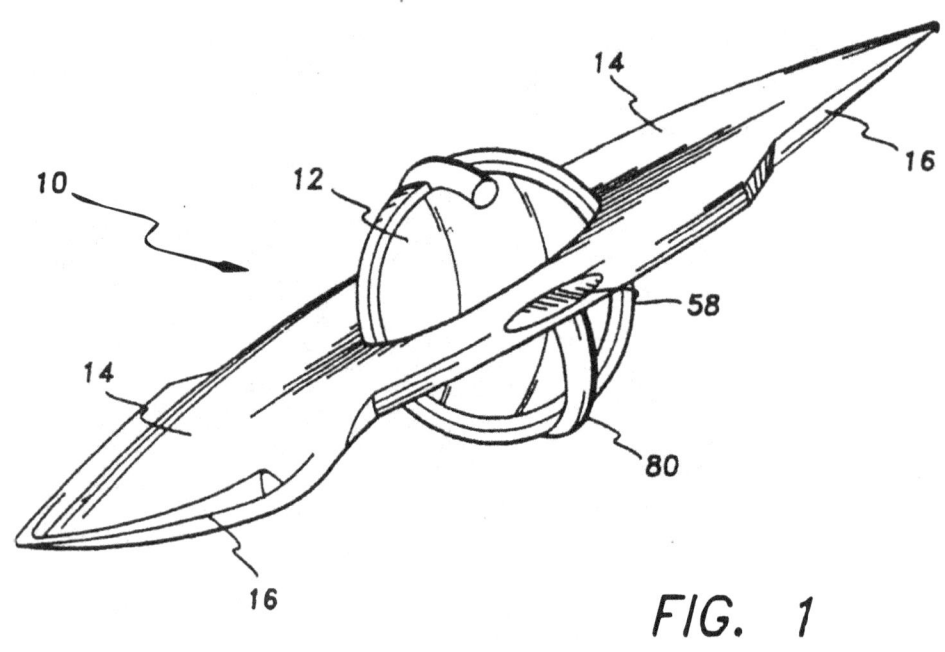

FIG. 1

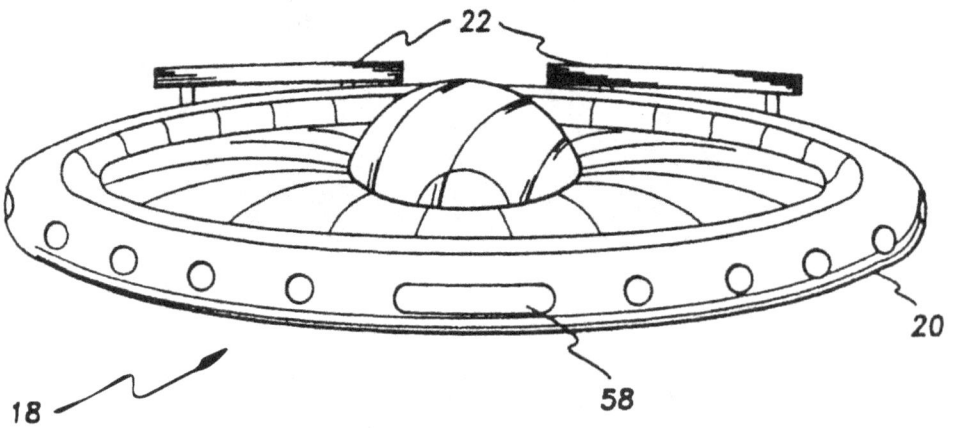

FIG. 2

United States Patent [19]

Webster

[11] **Patent Number:** **Des. 320,378**

[45] **Date of Patent:** ✶✶ **Oct. 1, 1991**

[54] **SPACECRAFT AIRCRAFT**

[76] Inventor: **Steven N. Webster**, 351 Zenith La., Juno Beach, Fla. 33408

[✶✶] Term: **14 Years**

[21] Appl. No.: **425,994**

[22] Filed: **Oct. 24, 1989**

[52] U.S. Cl. **D12/325**; D12/319

[58] Field of Search .... D12/319, 325, 330, 343; 244/23 C, 15, 52, 237

[56] **References Cited**

U.S. PATENT DOCUMENTS

D. 238,938	2/1976	Moller	D12/330
2,935,275	5/1960	Grayson	D12/325
3,022,963	2/1962	Frost et al.	244/15
3,243,146	3/1966	Clover	244/23 C
3,503,573	3/1970	Modesti	244/23 C
4,214,720	7/1980	DeSautel	244/23 C
4,901,948	2/1990	Panos	D12/343

Primary Examiner—Kay H. Chin
Attorney, Agent, or Firm—Richard C. Litman

[57] **CLAIM**

The ornamental design for a spacecraft/aircraft, as shown and described.

DESCRIPTION

FIG. 1 is a perspective view of a spacecraft/aircraft showing my new design;

FIG. 2 is a side elevational view thereof, the opposite side elevational view being a mirror image of that shown;

FIG. 3 is a front elevational view thereof;

FIG. 4 is a rear elevational view thereof;

FIG. 5 is a top plan view thereof;

FIG. 6 is a bottom plan view thereof;

FIG. 7 is a front perspective view thereof, with wing flaps lowered and landing gear retracted;

FIG. 8 is a front elevational view of a second embodiment of the spacecracft/aircraft;

FIG. 9 is a right side elevational view of FIG. 8, the left side elevational view being a mirror image of that shown;

FIG. 10 is a front elevational view of a third embodiment of the spacecraft/aircraft;

FIG. 11 is a right side elevational view of FIG. 10, the left side elevational view being a mirror image of that shown;

FIG. 12 is a front perspective view of FIG. 1, with flaps extended to provide a slow right turn;

FIG. 13 is a front perspective view of FIG. 1, with flaps extended to provide a sharp left turn;

FIG. 14 is a front and right side perspective view of the embodiment of FIGS. 8 and 9; and

FIG. 15 is a front and right side perspective view of the embodiment of FIGS. 10 and 11.

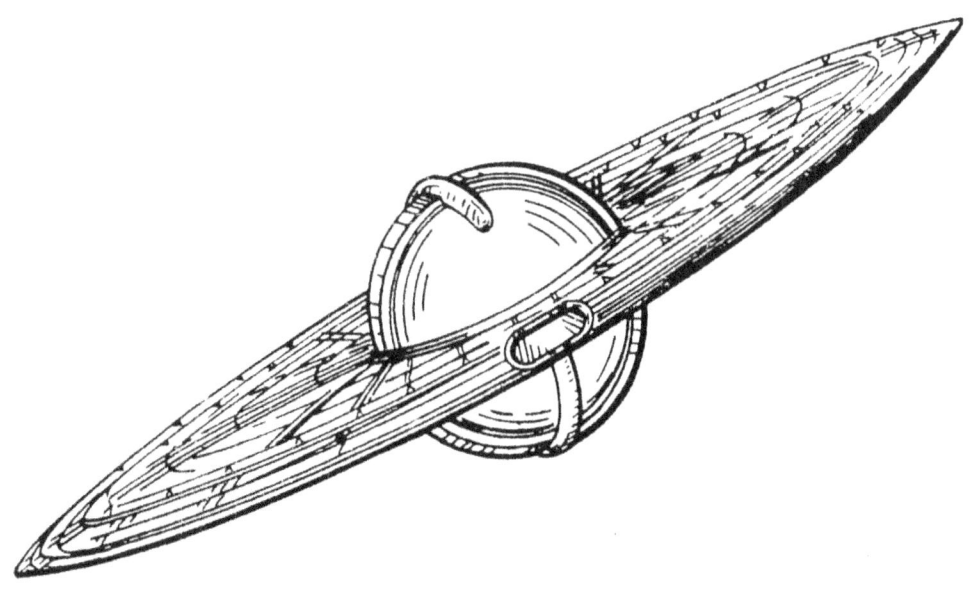

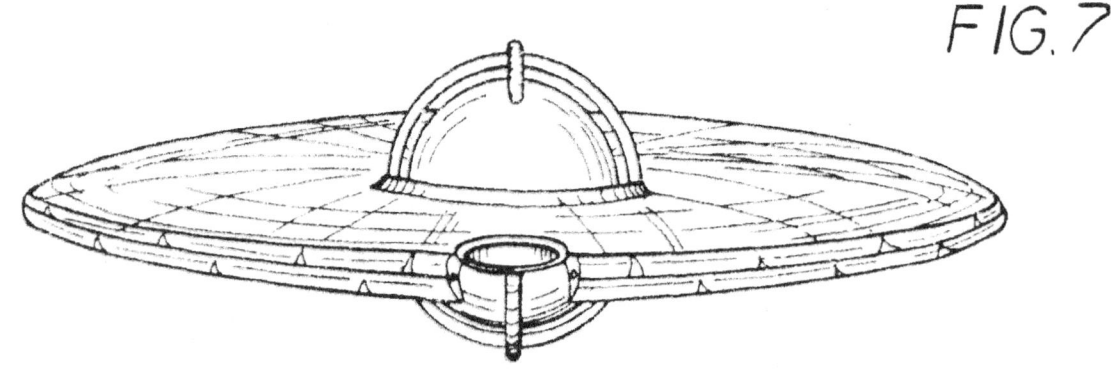

FIG. 7

Fig.12

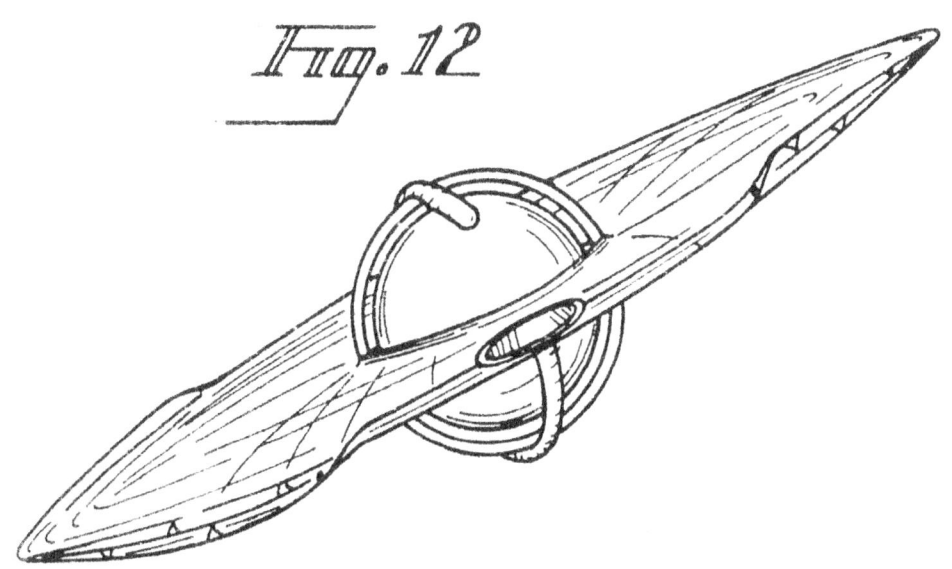

Fig.13

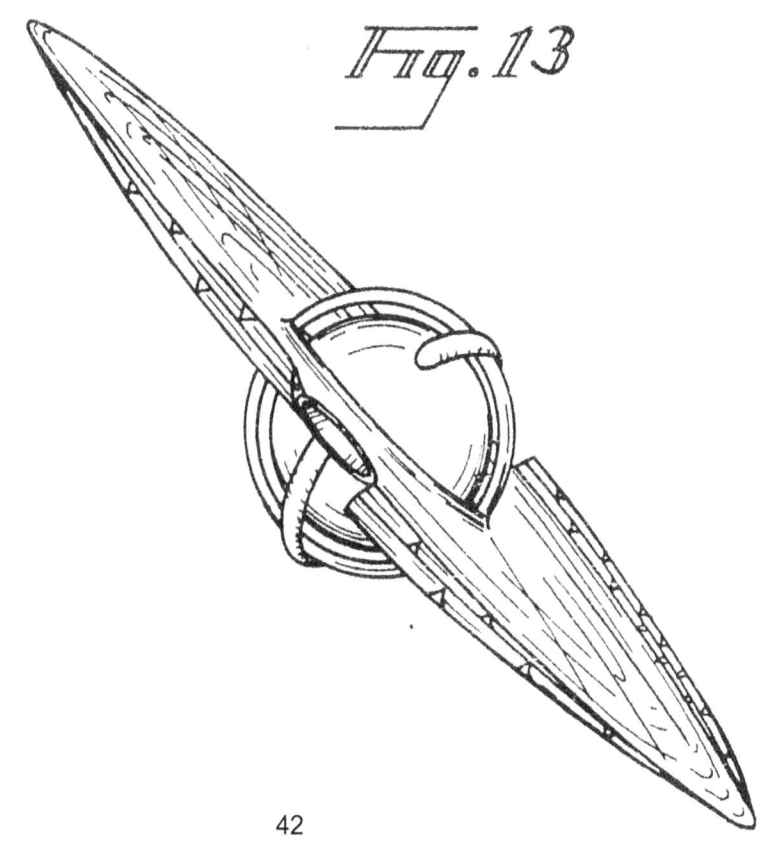

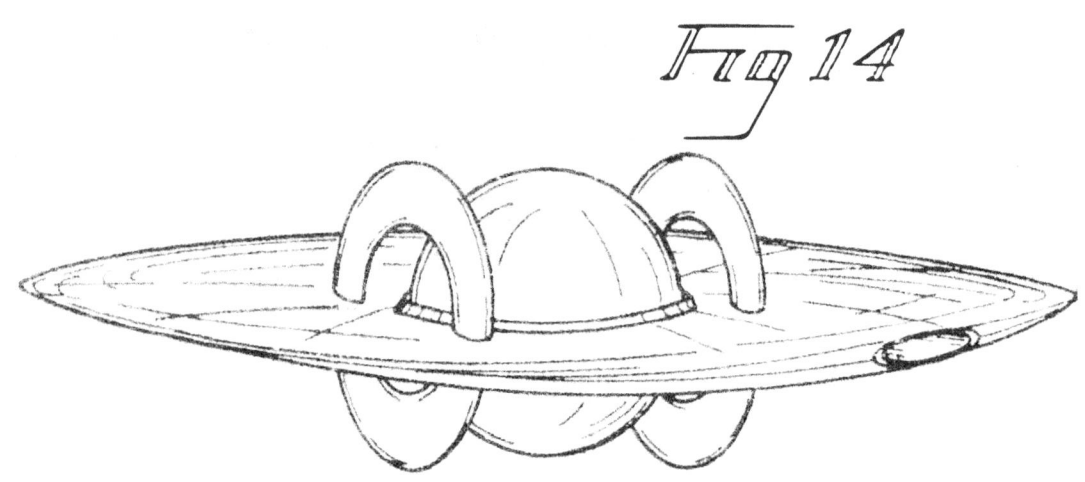

Fig 14

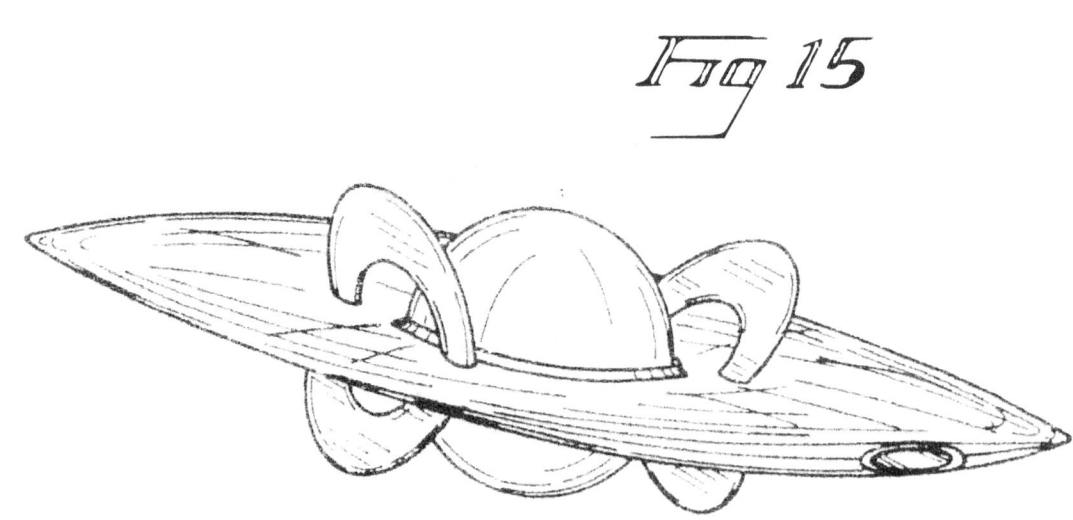

Fig 15

Patent Drawings Up-Date: Public: Project: Good Karma
USA Corporate/Government Funding Requested:
Contractor: Free Agent: Steven Nichols Webster
Drafted: August 028th, 2018
Up-Dated: September 01st, 2018
Team USA

FIG. 5-A-2

Earth's

Moon   Earth   Earth's Moon   Mars   Saturn   Jupiter   Venus   Sun

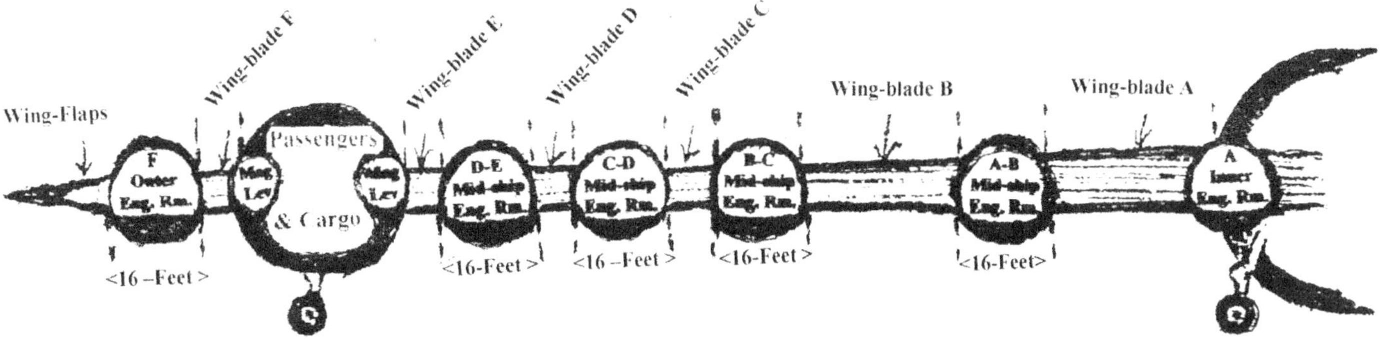

East   West

FIG. 5-A-2 was inspired this August 28th, 2018
looking south over the Atlantic Ocean from Atlantic Beach,
North Carolina. Our moon was bright to the East, then came
Mars, Saturn, Jupiter, Venus, all in clear view streaming
westerly. Point being; there are an endless amount of inspiring
observations to frame an aircraft/spacecraft after.
Endless: August 28th, 2018

September 25th, 2018

President Donald John Trump
The White House
1600 Pennsylvania Avenue
Washington, D. C. 20500

Steven Nichols Webster
400 Money Island Drive
Atlantic Beach, N. C., 28512
nickwebster1946@outlook.com
Ph: 970-946-3858

Reference: Authorization for NASA to develop the herein described Aeronautical Engineering Computer with a public education download.

The Honorable President of the United States of America
Donald Trump,

Sir, I ask you to compare the enclosed Project: "Good Karma" and Building Team USA and discuss this intended China, Russia, North Korea, NATO, et al "R&D" window with the Pentagon.

I ask you. Would it not be safer for our national defense to have NASA our United States Congress to carry on the world-wide invitation to develop this concentric aircraft/spacecraft "R&D" in the name and spirit of PEACE than for me alone to seek our neighboring ESA and Italian Space Agency to participate in completing my dreams?

Change of subject: Your United Nations speech this morning was very thoughtful and very pleasantly spoken from your heart.

Respectfully yours in Christ @ Sea & @ Home.
With PEACE of Mind!

Steven Nichols Webster
Steven Nichols Webster

Nick

45

Central Intelligence Agency
Office of Public Affairs
Washington. D. C. 20505

September 25th, 2018
Steven Nick Webster
Free Agent: SNW - Licensed
400 Money Island Drive
Atlantic Beach. North Carolina 28512
Nickwebster1946@outlook.com
Ph: 970-946-3858

Reference: I have sent this UPS presentation to both yourselves and the White House seeking the same overview on the same question of national security. For that overview to be observed from my perspective I have enclosed a copy of my said letter to Mr. President Trump herein.

Dear CIA,

Many years ago I asked both you and the FBI to check in on me from time to time because the cultural divide internal to our USA's social foundation was truly divided. The enclosed two paper-back books are for your review and assessment. I brought 20 copies each to distribute here in D. C. focusing on our U.S. Congress and our U.S. Senate.

In closing: Every known written language occurred during this waning of the Ice Age. We will see another atmospheric continuation of equal proportions appear from over our horizons or from within far more powerful than manmade weapons. This should inspire mankind to overcome the war laden history of our Earth's past.

Respectfully yours in Christ @ Sea & @ Home.
With PEACE of Mind !

Steven Nichols Webster

General Joseph F. Dunford Jr.
Chairman of the Joint Chief of Staff
1400 Defense Pentagon
Arlington, Virginia   20318

September 23, 2018
Nick Webster
400 Money Island Drive
Atlantic Beach, N. C. 28512

Reference: Attached two paper-back edition of Project: "Good Karma" and Building Team USA disclosing my public presentation of my here-by represented nuclear disarmament program seeking your approval for my continuing introduction of same to the European Space Agency, Italian Space Agency, and all countries as I am seeking to evolve technologically under one roof.

Sir,

I offer these writings; Project: "Good Karma" & Building Team USA, to fill the void of a quagmire arising from a; "Do as I say and NOT do as I did to get here.", dialogue while in the trenches of politics during nuclear disarmament procedures with a more reasonable dialogue. My purpose is to fill the void known to those peoples giving up their fight to carry their nuclear=warhead understanding while they decide. In the history of war; to die with honor, ment no surrender and was widely up-held. I say we; all countries, should discuss evolving technologically together for the remainder of eternity.

Respectfully yours in Christ @ Sea & @ Home.
With Peace of Mind !

Steven Nichols Webster
Steven Nichols Webster
Nick

Congressman Mac Thornberry – Texas   September 23, 2018
Chairman House Armed Services Committee   Nick Webster
2208 Rayburn House Office Building   400 Money Island Drive
Washington, D. C. 20515   Atlantic Beach, N. C. 28512
202-225-3706   970-946-3858

Honorable Chairman Mac Thornberry,

Attached are two paper-back editions of my life long attempt to
understand our United States history and how we could "Overcome
Eternal War"; 2017 & 2018. I will be in Washington D. C. but one more
day before returning to Atlantic Beach, N. C. As I write this
introduction to my Project: "Good Karma" and Building Team USA
I do not know if General Joseph F. Dunford Jr. has received his copies
as yet. Therefore, after your review of the attached two editions with
your Armed Services Committee I request your review be sent to
General Dunford, Chairman Joint Chief of Staff, Pentagon.

Respectfully Yours in Christ @ Sea & @ Home.
With Peace of Mind !

Steven Nichols Webster
Steven Nichols Webster
                    Nick

Congressman Lamar Smith - Texas

Chairman Science, Space, & Technology

2409 Rayburn House Office Building

Washington, D.C. 20512

202-225-4236

September 23, 2018

Nick Webster

400 Money Island Drive

Atlantic Beach, N.C. 28512

970-946-3858

The Honorable Congressman Lamar Smith
Chairman Science, Space, and Technology Committee,

Attached are 2-editions of Project: "Good Karma" indicating the progress and determination put forth on my part to become grant relative in regards to my North Carolina objective of receiving a grant for a N. C. university or college through Senator Rich Burr for an Aeronautical Engineering Computer to complete the student phase study of the "Great Circle" as presented therein. Therein, I ask you to present my objective of building Team USA to your Committee on Science, Space, and Technology.

Respectfully yours in Christ @ Sea & @ Home !
With Peace of Mind !

Steven Nichols Webster
Steven Nichols Webster

Nick

Chief Justice John G. Roberts, Jr.   September 23, 2018
United States Supreme Court   Nick Webster
#1 First Street N. E.   400 Money Island Drive
Washington, D. C. 20543   Atlantic Beach, N.C. 28512
                                                         970-946-3858

Your Honor Chief Justice Roberts,

Enclosed are my first two editions of my life-long study of the "Great
Circle". Why the Supreme Court ? Why now ? My option to question
and seek your opinion as to a citizen's right to represent an international
technology sharing concept towards "Overcoming Eternal War"; such as
mine, in the event that neither our Pentagon nor United States
Congress feel the same as I do.

Respectfully yours in Christ @ Sea & @ Home.
With Peace of Mind!

Steven Nichols Webster
Steven Nichols Webster
          Nick

2 paperbacks enclosed; Project: "Good Karma" 2017
                    & Building Team USA 2018

Building Team USA has been brought to you by

Webster's
Home-Schooling

Two Boats School

Grades K to Captain

A nautical approach to introducing math
with a Christian background
and Captain's license
leadership outlook for
God and Country

Textbooks available @ Barns & Noble and Amazon

Copyrights:

Listen up !

No
Monkey Business
In the
Wheelhouse

Ok !

That also means no food or drinks on the chart table.
No food or drinks of any kind on or near electronics.

Anyone with children, these last pages are for you.
Here is the beginning of our Great Circle Study

Dear Pre-Readers,

The audible joy of being read to has
been known for millenniums.
A millennium is 1,000 years.
A century is 100 years.
A decade is 10 years.
And, how old are you ?
I am_____ years old.
My name is_____.

You that can read, read to a pre-reader.
Read to a pre-reader as a big brother.
Read to a pre-reader as a big sister.
Read to a pre-reader as a friend.

Welcome aboard !

The

Sea of Math

"Division"

An Introduction to Time and the four Directions

North, East, South, and West

Benchmarks Achieved

An Introduction to

"*Division*"

1/2s, 1/3s, 1/4s, 1/5s, & 1/6s

Grades: K~3

<u>Learning how to tell time.</u>
How mathematical time came to be;
60 seconds a minute, 60 minutes an hour, 24 hours a day.

<u>Learning our four directions; North, East, South, and West.</u>
How the 360 Degrees of our compass
came to be.

*Our highest academic benchmark
achieved in this lesson is
Division.*

The ship sailed by students "The Sea of Math" is found on an
exterior wall of the Pagosa Springs Elementary School in
Archuleta County School District, Colorado.

This is a

Great Circle Study

By

Chick

First we are going to learn something about

Mathematical Compatibility.

*We will start by answering the question: Why do we have
12 hours, 60 minutes, and 60 seconds on the face of our clock?*

*We will then take this observation one step farther to answer the
question: Why do we have 360 degrees to our nautical compass?*

We will start with a circle.

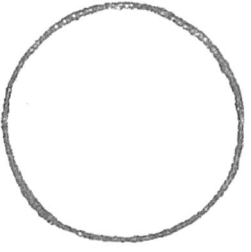

Now, we divide that circle in half.

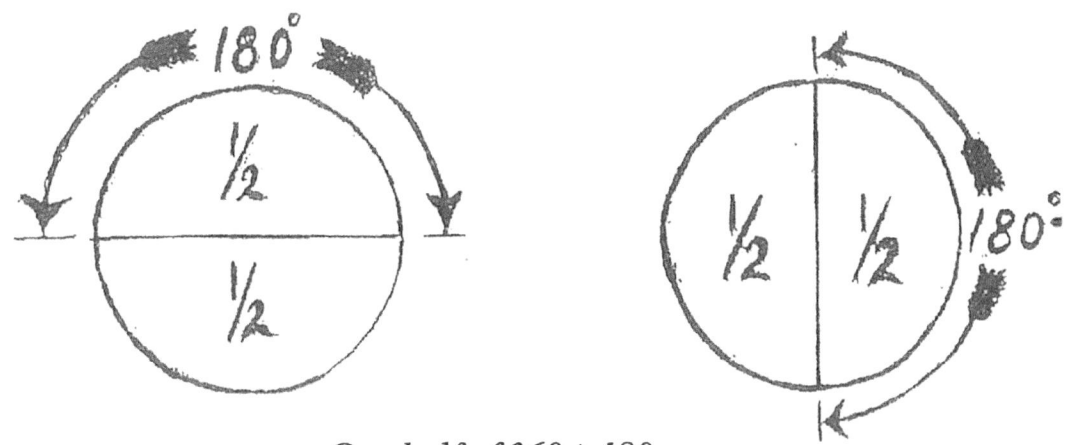

*One half of 360 is 180.
We already know there are 360 degrees in a circle.
We are about to learn why.*

$$\frac{1}{2} \times \frac{1}{2} = \frac{1}{4}$$

One half of one half is one quarter.

One quarter of 360 is 90.

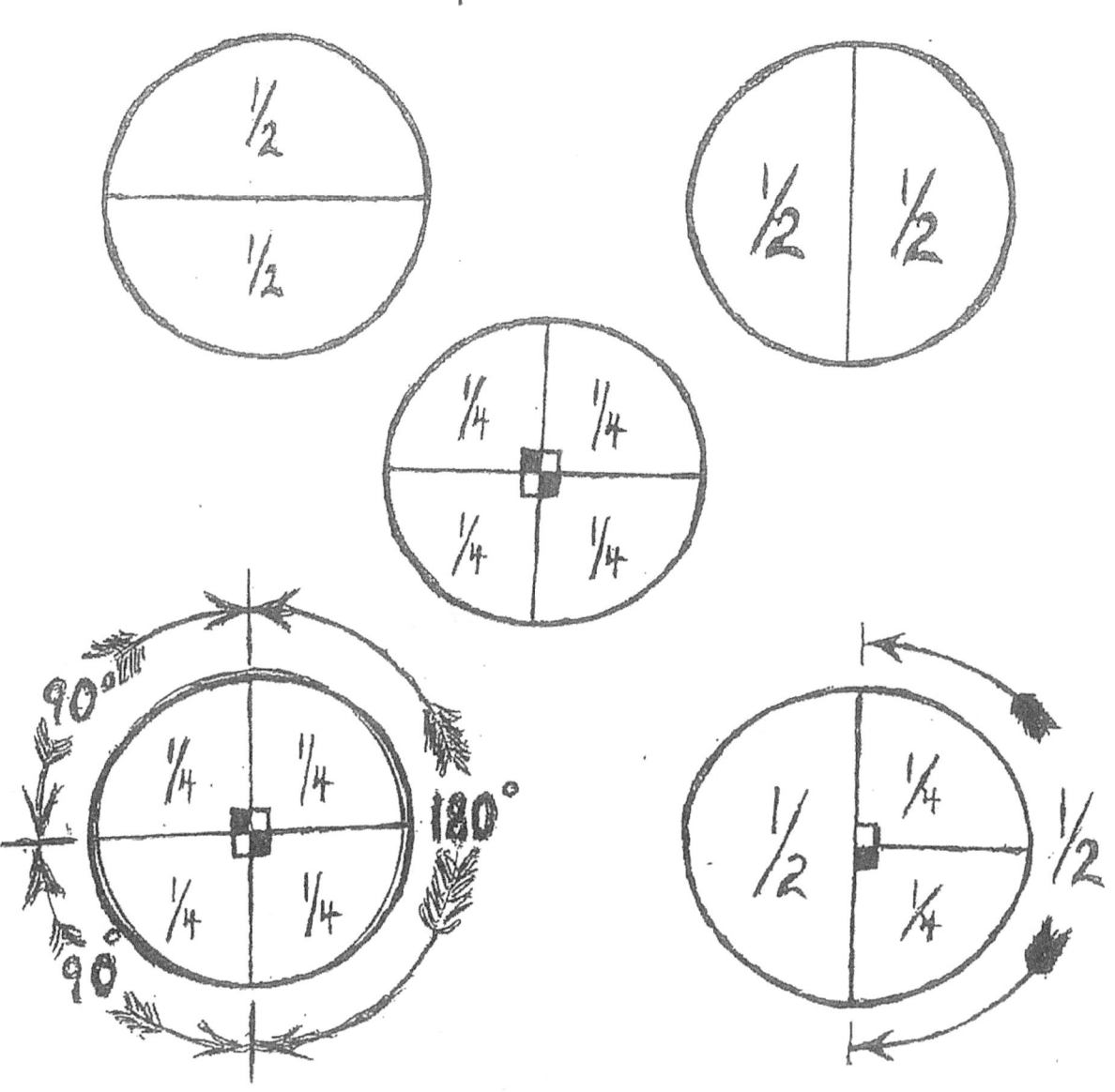

1/3 x ¼ = 1/12

One third of one quarter equals one twelfth.
The 12 hours on the face of our clock.
The hour hand goes around twice for the
12 hours of day and 12 hours of night.
A. M. means At Morning.   P. M. means Past Morning
Midnight to Noon   Noon to Midnight

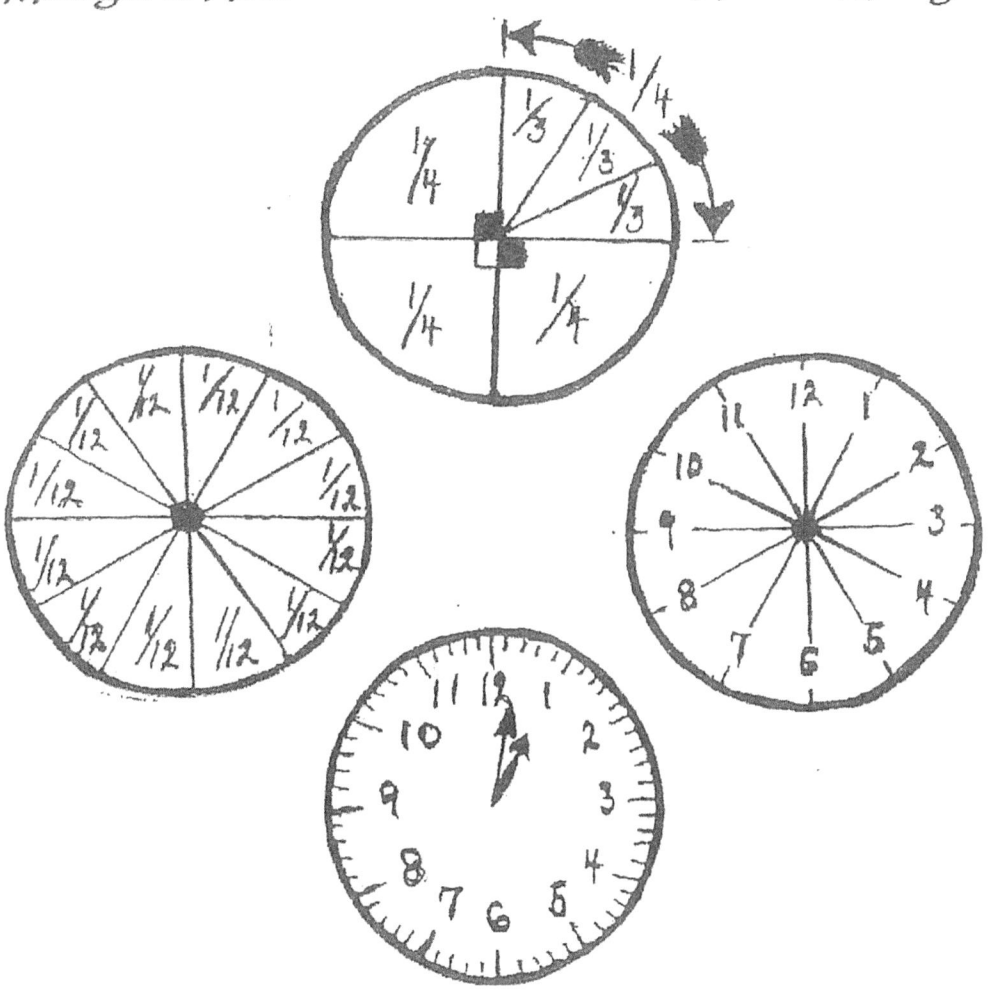

8

1/5 x 1/12 = 1/60

One fifth of one twelfth equals one sixtieth.
The 60 seconds of every minute and the 60 minutes of every hour.

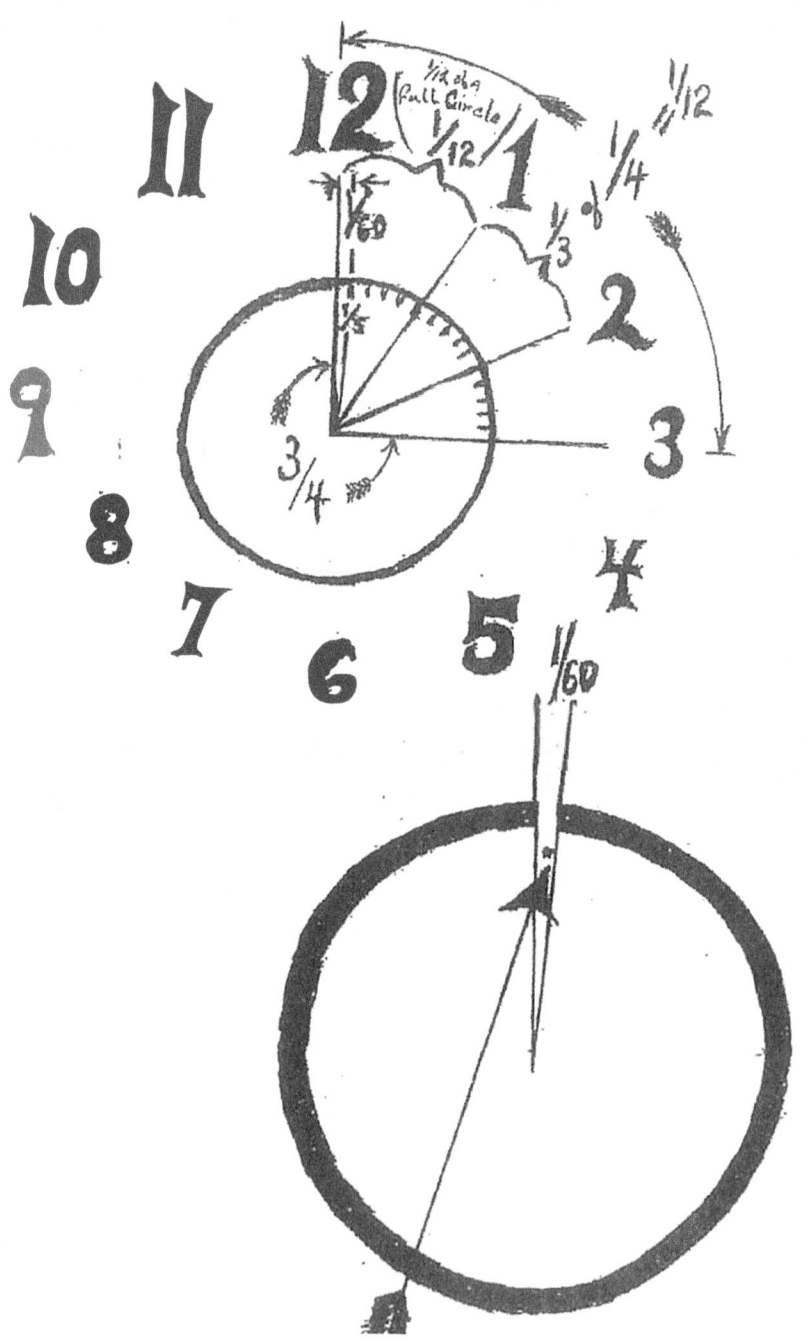

1/6 x 1/60 = 1/360

One sixth of one sixtieth equals one three hundred sixtieth.
This gives us the 360 Degrees of our Nautical Compass.

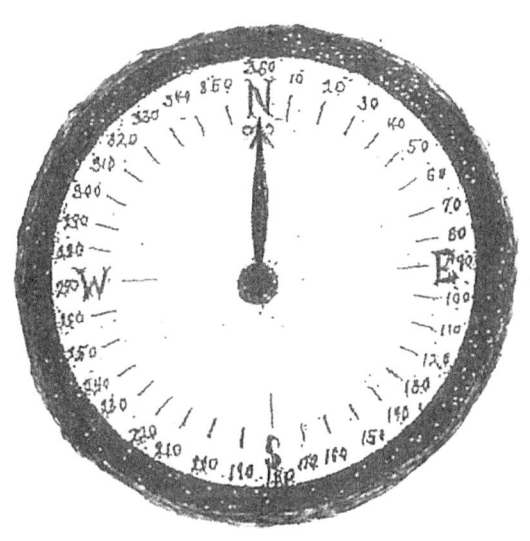

&

The 360 Degree CIRCLE

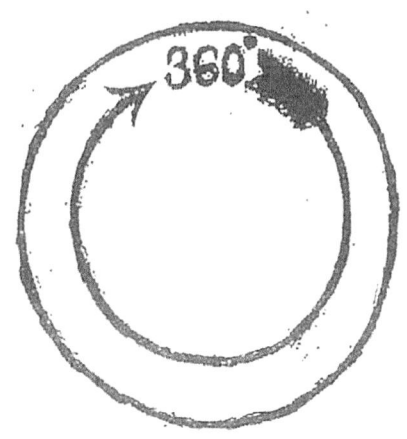

Latitudes

Division

We divided a circle by 1/2.
The we divided that 1/2 into 1/4s.

That is how latitudes took form.

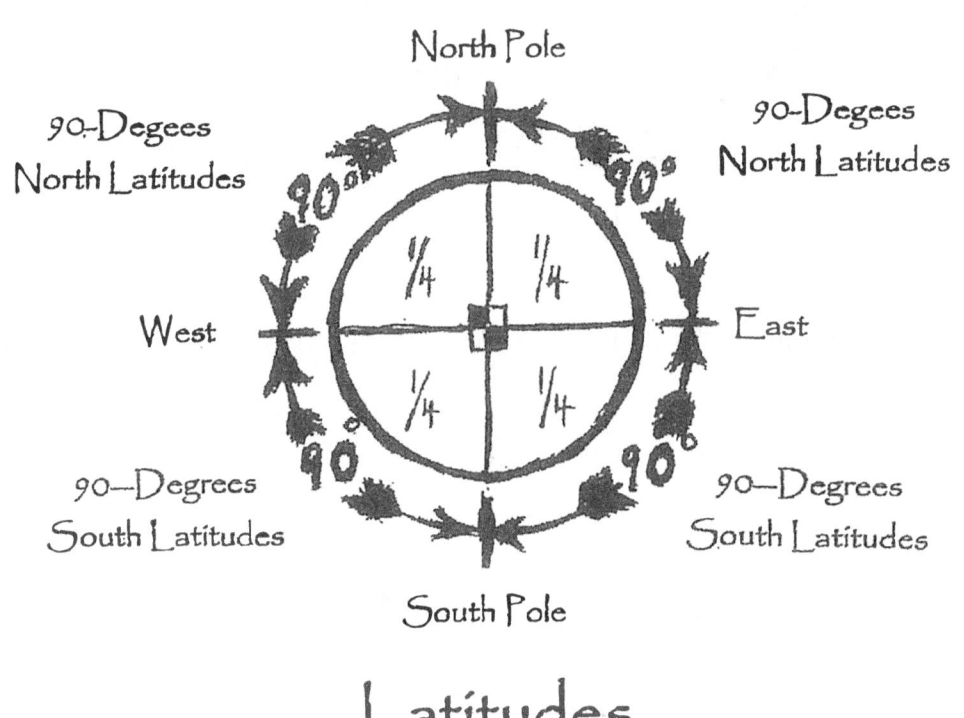

Latitudes

Longitudes

Division

We divided a circle by 1/2.

That is how longitudes took form.

180-Degrees West Longitudes

180-Degrees East Longitudes

Longitudes

Our subject of focus is now the face of a Clock or Watch

"Hi !" I am your talking clock.

My name is Tick-Talk.

I have three hands to tell time.
That is all I do. I tell time.

#1:

I am your Hour hand.
I point to the hour; 24 hours a day,
12 hours of day, and 12 hours of night.
I am pointing to a little past One O'clock.

#2:

I am your Minute hand.
I point to minutes, 60 minutes an hour.
I am pointing to Ten Minutes after One O'clock.

#3:

I am your Second hand.
I point to the seconds, 60 seconds a minute.
I am pointing to Forty Seconds after Ten past One O'clock.
However, you will not be tested on this, your Second hand.

Telling Time Test

Student's Name: _____

What time is it?

Date: _____

Grade in school: _____
Pre-study test grade: _____
After-study test grade: _____

#1:

The time is _____.

#2:

The time is _____.

#3:

The time is _____.

#4:

The time is _____.

#5:

The time is _____.

#6:

The time is _____.

#7:

The time is _____.

#8:

The time is _____.

#9:

The time is _____.

#10:

The time is _____.

Telling Time Test

Student's Name: _____

What time is it?

Date: _____

Answers

Grade in school: _____
Pre-study test grade: _____
After-study test grade: _____

#1:

The time is __01:00__.

#2: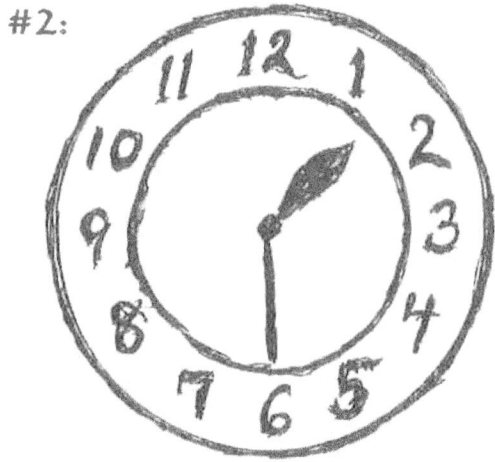

The time is __01:30__.

#3:

The time is __07:20__.

#4:

The time is __07:00__.

Answers

#5:

The time is __03:00__.

#6:

The time is __03:45__.

#7:

The time is __10:00__.

#8:

The time is __06:05__.

#9:

The time is __05:40__.

#10:

The time is __12:00__.

According to Captain Soh Cah Toa the Sea of Math
gets rough at times.

Lesson #1

OFFICE

Capt. Soh Cah Toa

OK, I'm knocking.

Hey, I can see eye to eye
with the keyhole.

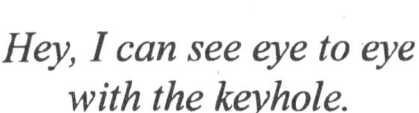

Never mind the keyhole.
Do you want to go in and study or not?

Who is Captain Soh Cah Toa?

Captain Soh Cah Toa is actually a code to the Navigation Triangle.

The Navigation Triangle has three sides.

Easy as 1, 2, 3, and A, B, C.

#1:   A = Latitude
#2:   B = Longitude
#3:   C = Our Course

Administrators, parents and teachers know the Captain as an acronym.
An interesting acronym wherein the first letter of each name
refers us to a function of trigonometry and the
next two letters to a side of the triangle.
Give your children a head start.
Give each child a complete
mathematical overview
of "Navigation" in
the beginning.
Grades K- 12

The Navigation Triangle is designed with Mathematical Compatibility

The Two Circles of Time

Genesis 1:1 "In the beginning God created the heavens and the earth."
Evolution preceded the written word; athletic records are bettered all the time.

The 1ˢᵗ Circle of Time;
God *made the*
1ˢᵗ Circle of Time.

The 2ⁿᵈ Circle of Time;
Mankind *made the*
2ⁿᵈ Circle of Time.

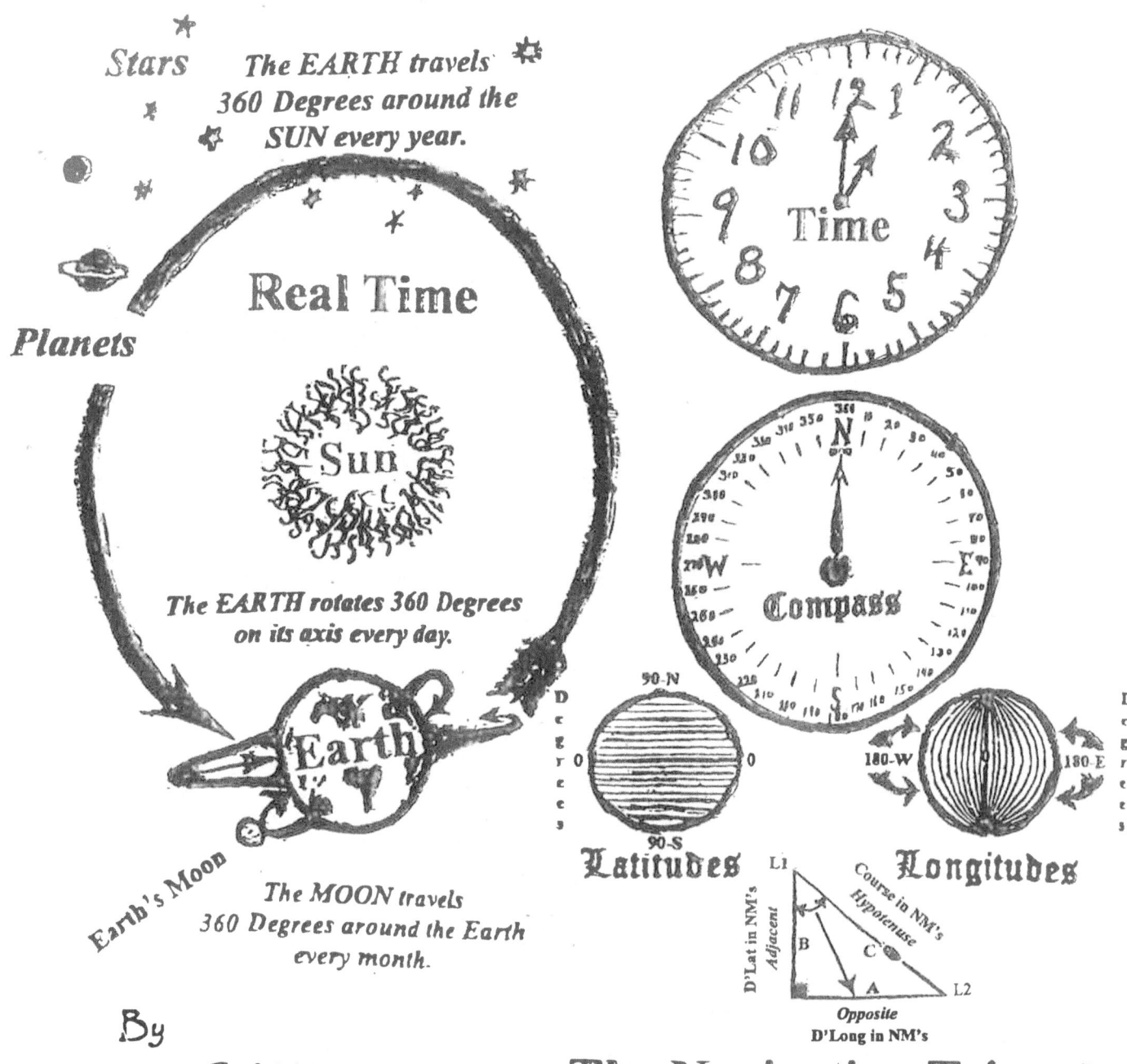

Stars

The EARTH travels 360 Degrees around the SUN every year.

Real Time

Planets

Sun

The EARTH rotates 360 Degrees on its axis every day.

Earth

Earth's Moon

The MOON travels 360 Degrees around the Earth every month.

By

S. N. Webster

Time

Compass

90-N
Degrees
0   0
90-S
Latitudes

180-W   180-E
Degrees
Longitudes

L1
Course in NM's
Hypotenuse
D'Lat in NM's
Adjacent
B
C
A
L2
Opposite
D'Long in NM's

The Navigation Triangle

The 2ⁿᵈ Circle of Time follows the 1ˢᵗ Circle of Time.
Wherein we study the Navigation Triangle in Lesson #1.

Welcome Aboard

A, B, & C go to Sea;
easy as 1, 2, 3 !

A = Latitude

B = Longitude

C = Course

Latitudes  , **Longitudes** 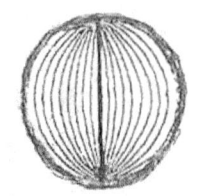 ,

and our

Course

Who wants to get an "A" for today? Come on raise your hand. Everybody who wants to get an "A" today raise your hand.

Easy as
1, 2, 3, & A, B, C

OK! To get an "A" today you must remember 3-things.

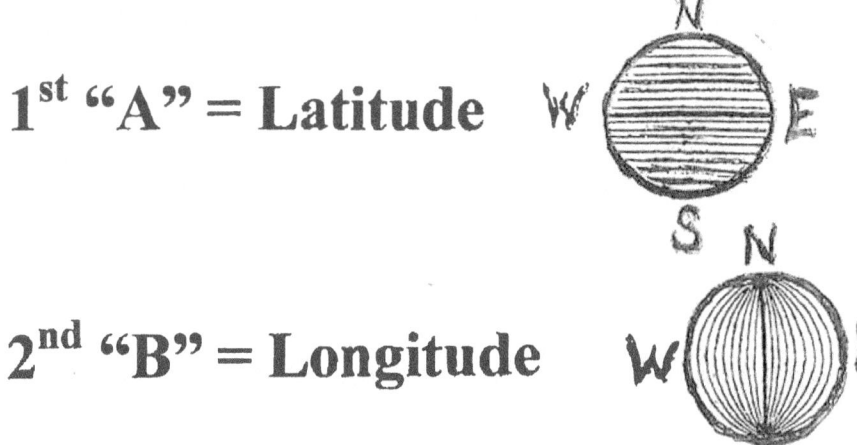

1ˢᵗ "A" = Latitude

2ⁿᵈ "B" = Longitude

3ʳᵈ "C" = Our Course line

A = 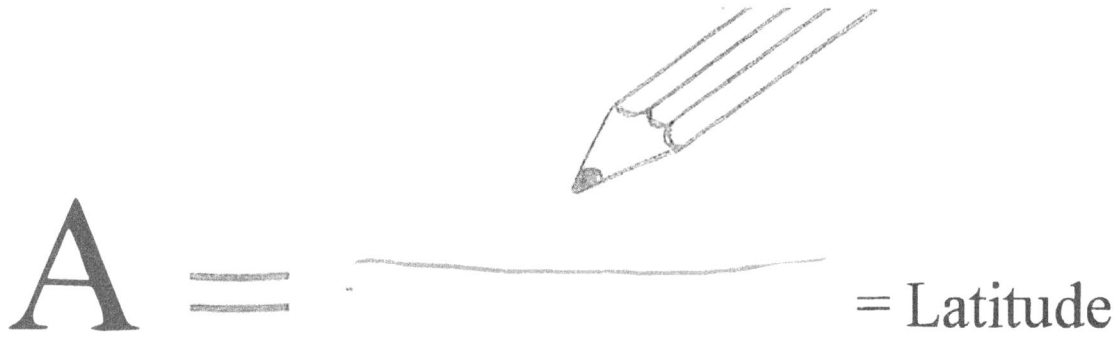 = Latitude

Remember the crossbar of the A. That may help you remember what Latitudes look like as they stretch East and West. Latitudes are Straight Horizontal Lines on CHARTS.

A ⸗ = Latitude

Now draw a Latitude next to the A.

A = = Latitude

To get an "A" today draw a Latitude next to the A above.

Latitudes are Straight Horizontal Lines on CHARTS.

Tomorrow you will be asked to remember A = Latitude. Latitudes are Straight Horizontal Lines that stretch East & West on CHARTS.

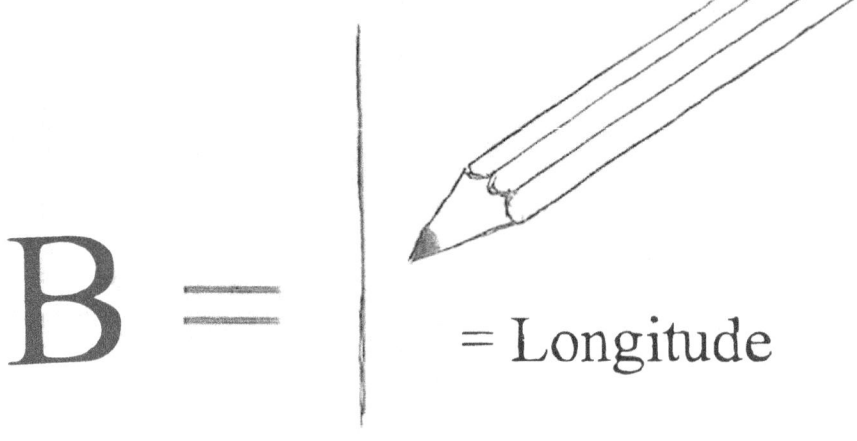

B = | = Longitude

Remember the backbone of the B. That may help you remember what Longitudes look like as they stretch North and South. Longitudes are straight vertical lines on CHARTS.

B = = Longitude

Now draw a Longitude next to the B.

B = = Longitude

To get an "A" today draw a Longitude next to the B above.

Longitudes are Straight Vertical Lines on CHARTS

Tomorrow you will be asked to remember B = Longitude. Longitudes are Straight Vertical Lines that stretch North and South on CHARTS.

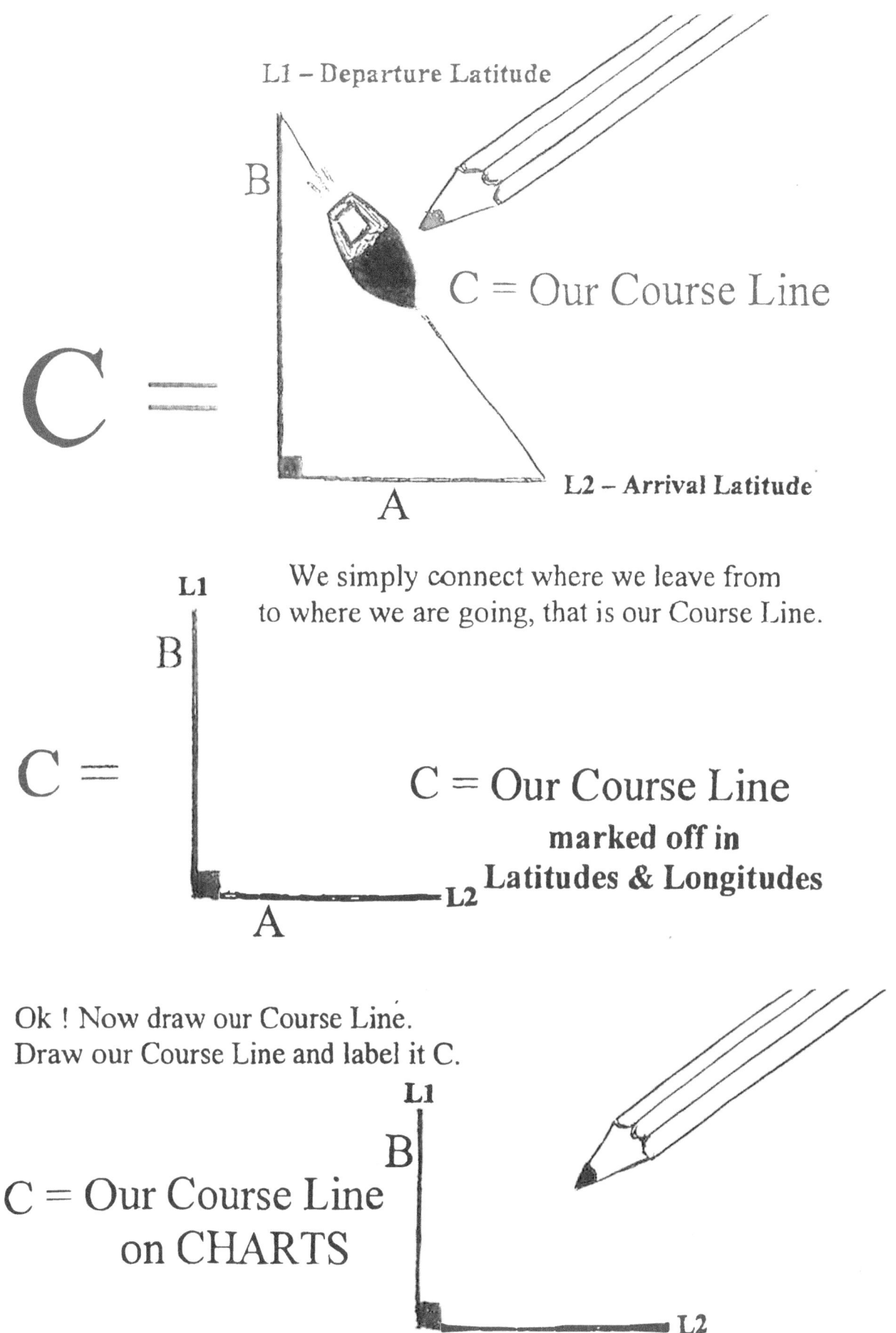

L1 – Departure Latitude

B

C = Our Course Line

C =

L2 – Arrival Latitude

A

L1

B

We simply connect where we leave from
to where we are going, that is our Course Line.

C =

C = Our Course Line
marked off in
Latitudes & Longitudes

A

L2

Ok ! Now draw our Course Line.
Draw our Course Line and label it C.

L1

B

C = Our Course Line
on CHARTS

A

L2

Climbing the
LADDER of
Latitudes

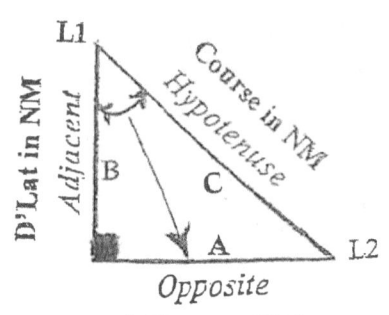

L1

D'Lat in NM
Adjacent

B

Course in NM
Hypotenuse

C

A

Opposite

L2

D'Long in NM

Cos Mid-Lat x D'Long in Minutes = Nautical Miles
Page #581: American Practical Navigator, Volume #2

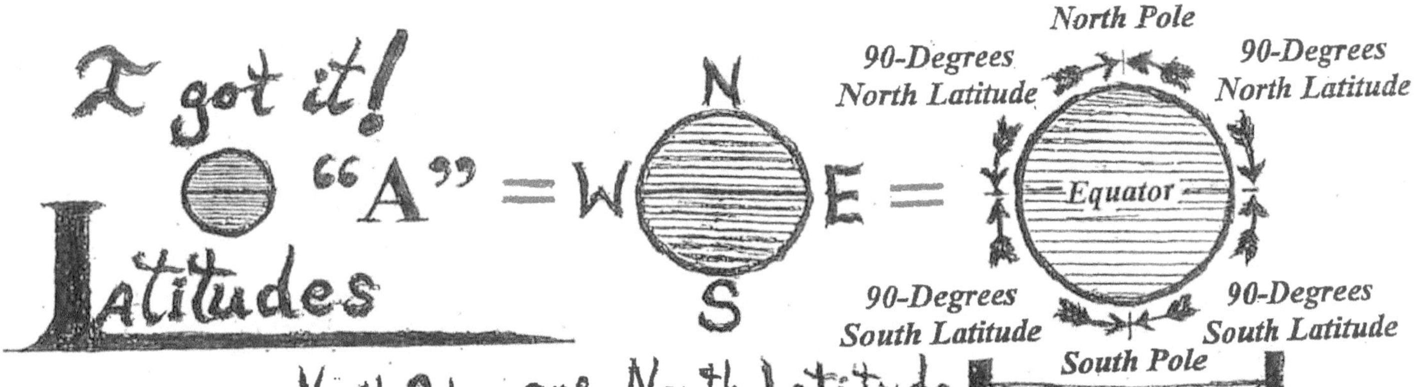

I got it!

"A" = W ⊙ E =

N

S

North Pole

90-Degrees
North Latitude

90-Degrees
North Latitude

Equator

90-Degrees
South Latitude

90-Degrees
South Latitude

South Pole

Latitudes

North Pole - 90° - North Latitude
80° - North Lat.
70° - North Lat.
60° - North Lat.
50° - North Lat.
40° - North Lat.
30° - North Lat.
20° - North Lat.
10° - North Lat.
0 - Equator
10° - South Lat.
20° - South Lat.
30° - South Lat.
40° - South Lat.
50° - South Lat.
60° - South Lat.
70° - South Lat.
80° - South Lat.
South Pole 90° - South Latitude

Latitudes

Now, stand up straight like a longitude.
Stand up straight from head to toe like a longitude.

You remember "A" our Latitudes.
Well, use the Equator for your belt
and stand tall like a Longitude.

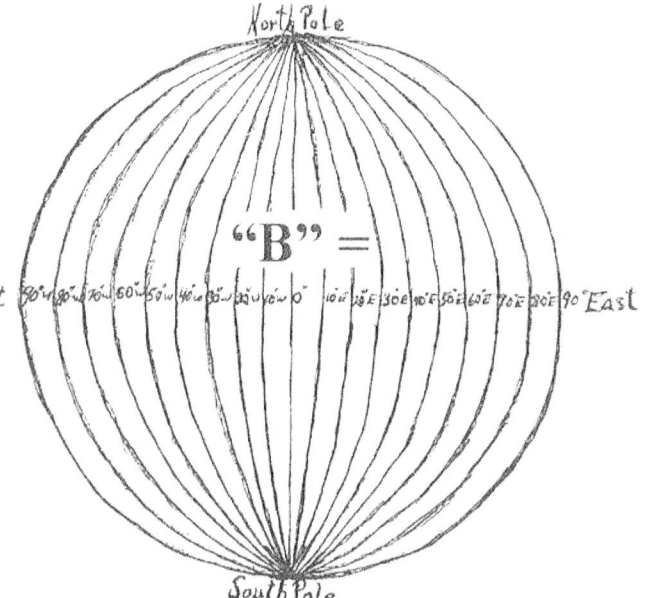

"B" =

180-Degrees West Longitude and 180-Degrees East Longitude
Taveuni, Fiji
International Date Line

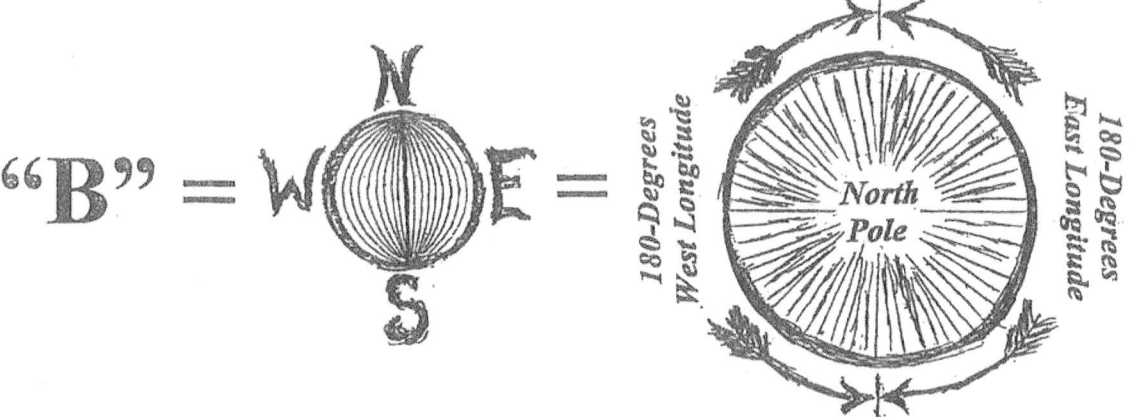

Zero-Degrees
Longitude
Greenwich, England
Zulu
Prime Meridian

Longitudes

Again, stand up straight like a longitude.
Stand up straight from head to toe like a longitude.

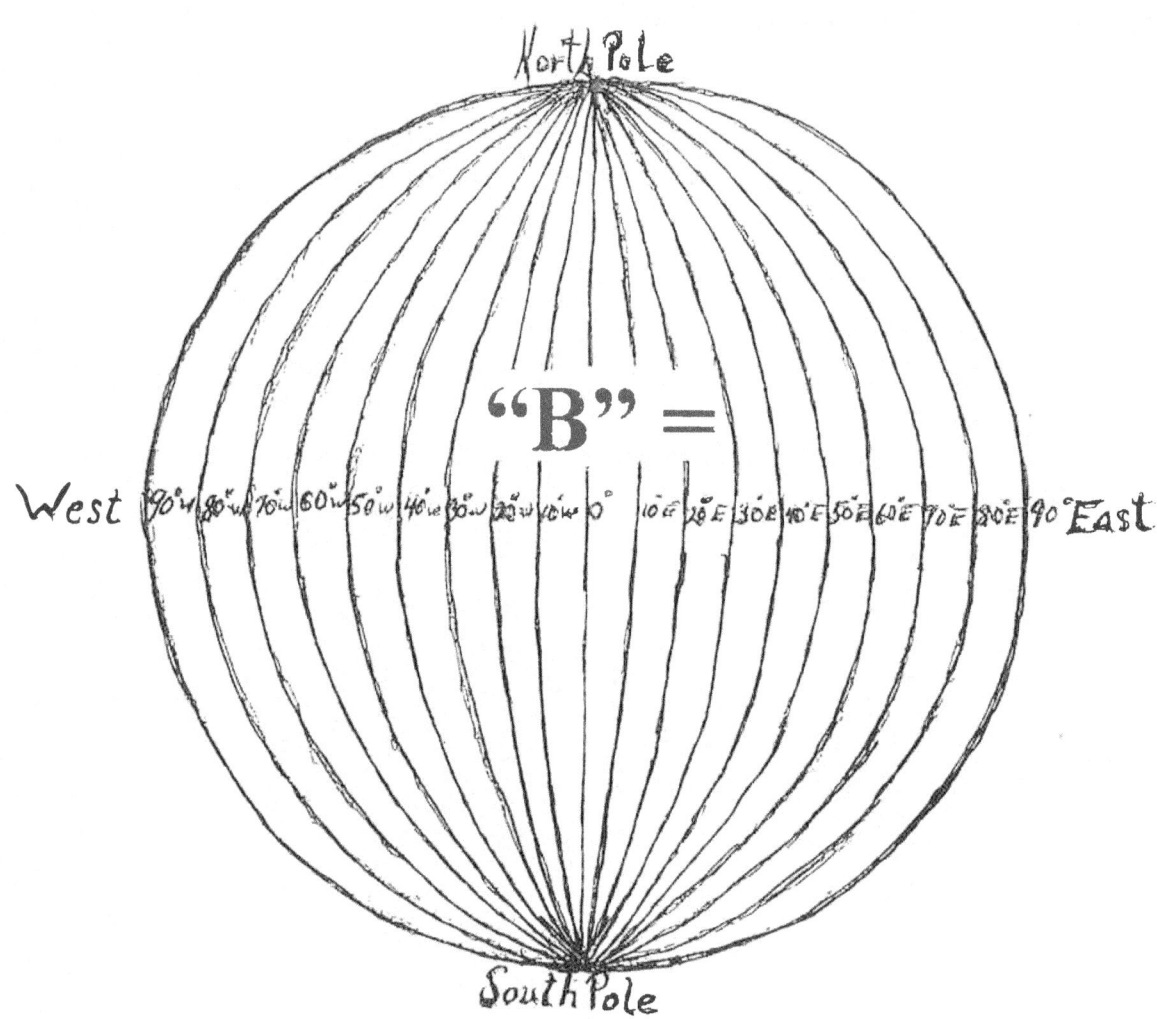

Longitudes

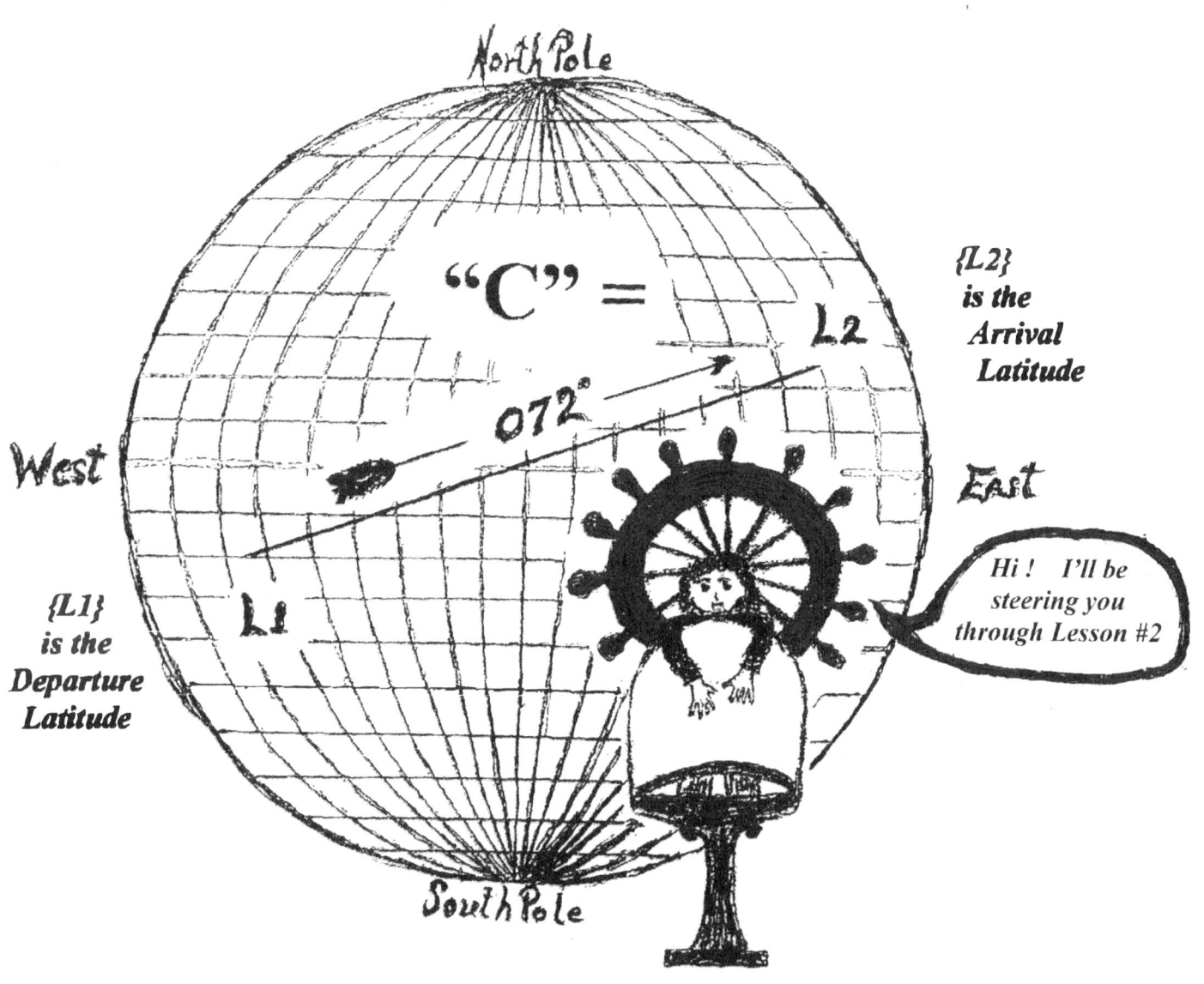

Our Course is named by the
360-Degrees of our Nautical Compass.

This particular Course Line is 072-Degrees.

C-072°

Course

It is really a small thing when you look at it.

Grades K-3

Academic Checklist

Know your home address. ☑

Know your home phone number. ☑

Know your teacher's name. ☑

Know what time it is. ☑

Know what direction you face
when you enter and leave your ☑
home, as the sun rises in the east
and sets in the west.

OK! Let's get ready for the test on the following pages.

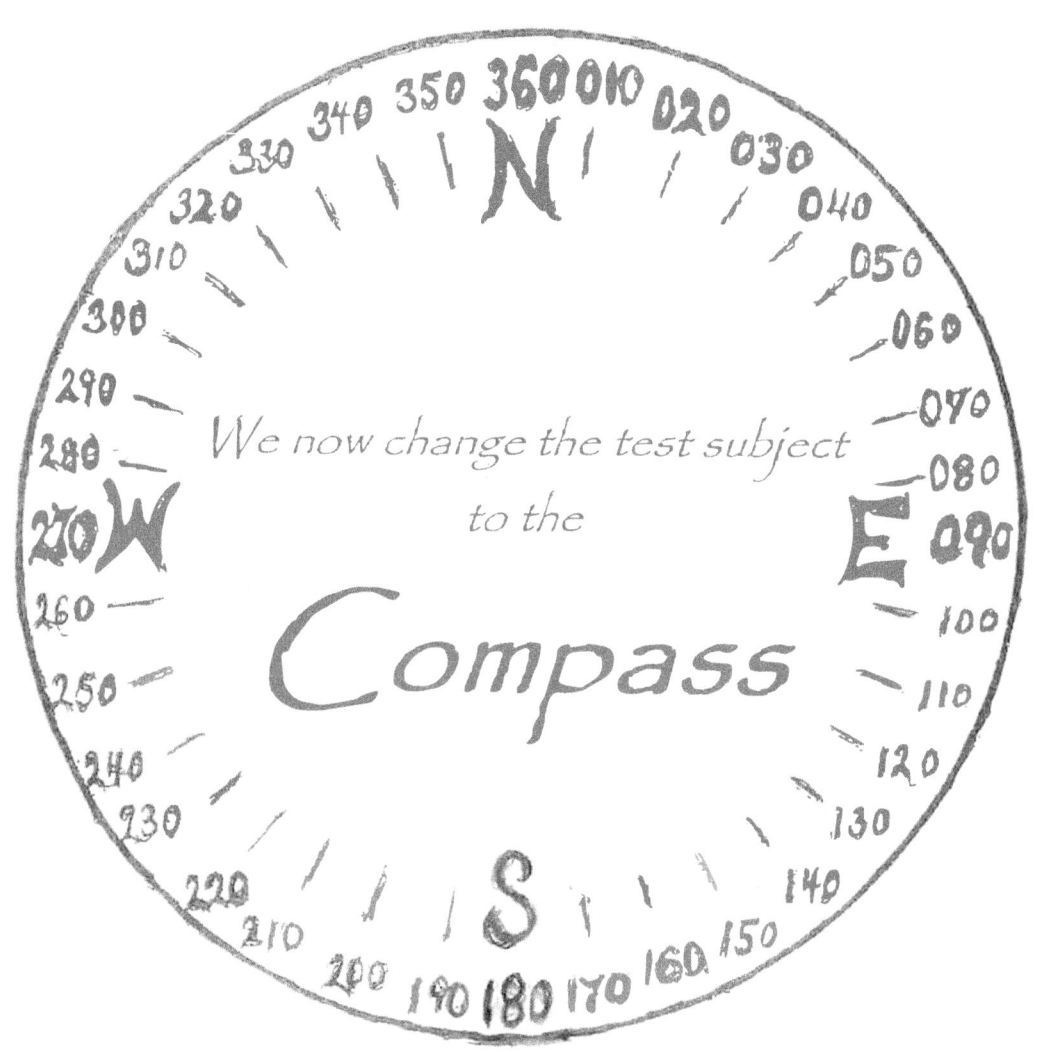

We now change the test subject

to the

Compass

The
Nautical Compass

The 4 Cardinal Directions: North, East, South, & West
&
The 4 Quadranrts of Compass Notation:
North East, South East, South West, & North West

North

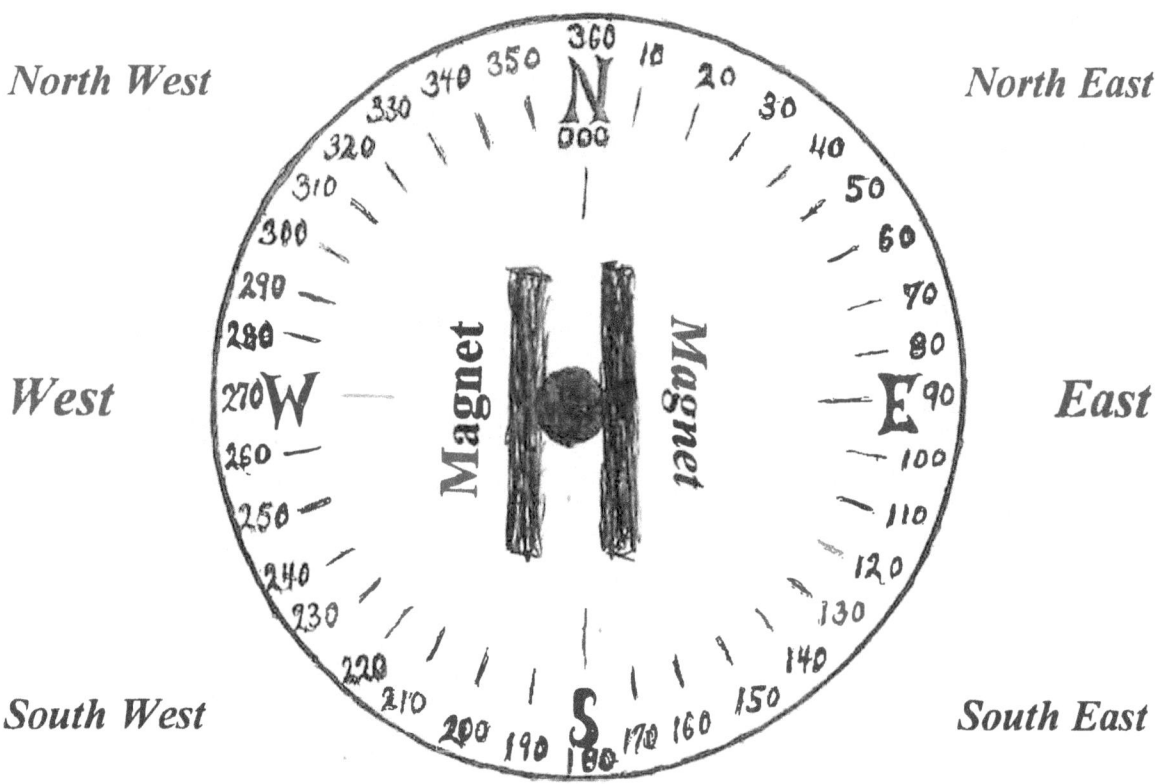

North West

North East

West

East

South West

South East

South

Compass Headings

and

Compass Bearings

Compass Vocabulary

and

Discussion

Grades 6-12-Captain

Webster's Vest Pocket Dictionary
gives the following explanation for
{Direction}.

Direction: 1: supervision
2: order
3: course along which something moves

We all know that a compass points to the north. Magnetic north attracts iron. Many say magnetic north was formed millions of years ago when meteorites imbedded the Great Lakes area that now joins Canada and our United States. Today it is as if magnetic north were an offsetting longitude that affects compasses in varying strengths and in differing; yet, controllable situations. Because we have studied the errors encountered over the years while using a compass we have come to call these errors either Variation or Deviation. We also name the error either East or West.

Compass Vocabulary: Grades: 6-12- Captain

... indicates that only a portion of the original explanation is being used.

Direction: "Any of the 360 degrees encompassing North, East, South, and West." snw ... Initially, compasses were used only to indicate north... The directions were given the names of the various winds, now known as North, East, South, and West; these are the cardinal directions... Modern compasses use the standard 360-degree system...
   Dutton's Nautical Navigation, 15th Edition, Chapter #7, Page 69, Article #701

Bearing: "The direction; compass reading, taken from you; your vessel, towards an object of interest at a particular moment expressed in degrees 000 - 360." snw ... the navigator must be able to measure and express the direction of these things; objects ashore or aids to navigation...
   Dutton's Nautical Navigation, 15th Edition, Chapter #1, Page #5, Article #109

In our test questions to come; Vessels A, B, C, & D are
aligned with the center of our compass rose, outgoing,
and their bearing and direction are the same.
The bearing and direction for Vessels E & F are not the same.

Heading: The direction in which a ship points or heads at any instant, expressed in angular units, 000-degrees clockwise through 360-degrees
   Dutton's Nautical Navigation, 15th Edition, Chapter #1, Page #4, Article #109

Course: ... In other words, "course" is your intended direction; while "heading" is the actual direction you are steering at any given instant.
   Dutton's Nautical Navigation, 15th Edition, Chapter #1, Page 4, Article #109

Students,

The following test pages depict a fleet of friends leaving a temporary anchorage in the Bahamas. Your vessel is in the center of those vessels and in the center of the anchorage. You are simply recording an observation for a journal as your friends depart for the Jupiter Inlet Light at different times and with different voyage plans.

For this test simply check the time, heading, and bearing for each of your friends; Vessel A, Vessel B, Vessel C, Vessel D, Vessel E, Vessel F.

In this particular situation the bearing and heading of the first four vessels of interest will be the same. We will use Protractor Triangles, ParaLock Plotter, or a conventional parallel rule; if you have one, as we discern the headings of Vessels E & F.

Compass Test
North, East, South, & West
"Headings" and "Bearings"

What is the heading on _Vessel A_ ?

_____

What is the bearing on _Vessel A_ ?

_____

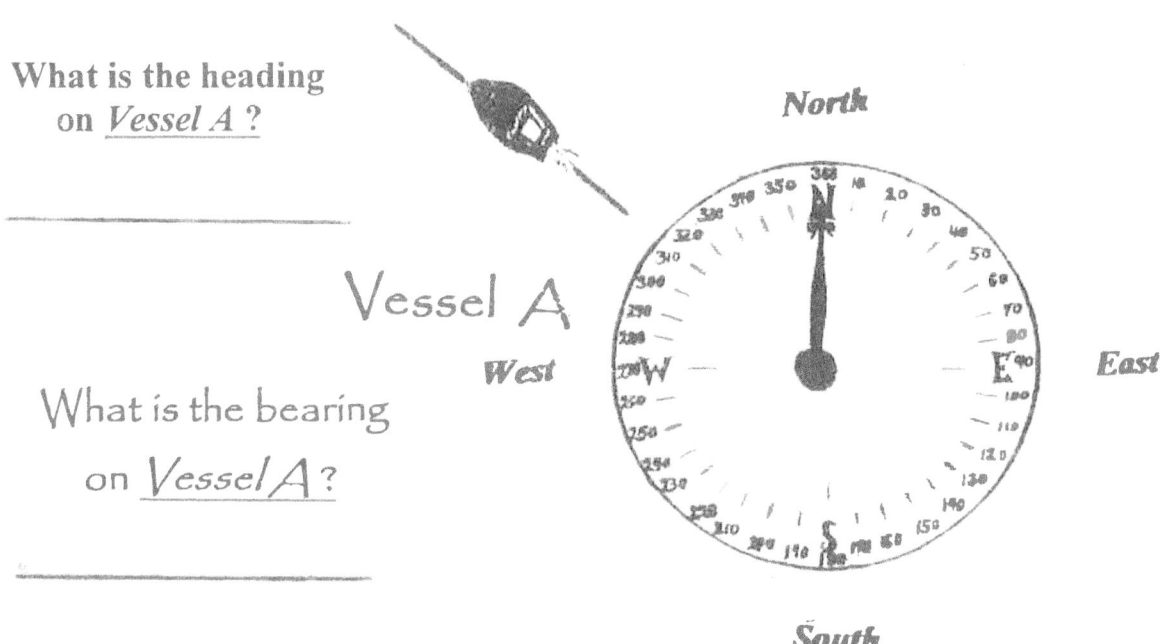

Vessel A

Vessel B

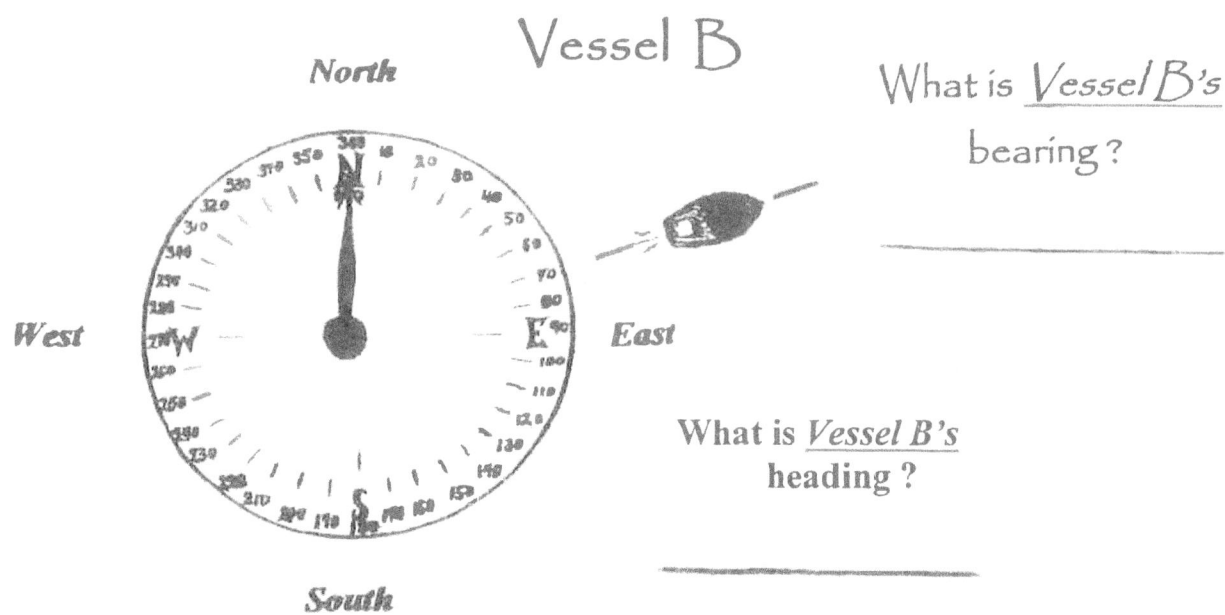

What is _Vessel B's_ bearing?

_____

What is _Vessel B's_ heading?

_____

Compass Test
North, East, South, & West
"Headings" and _"Bearings"_

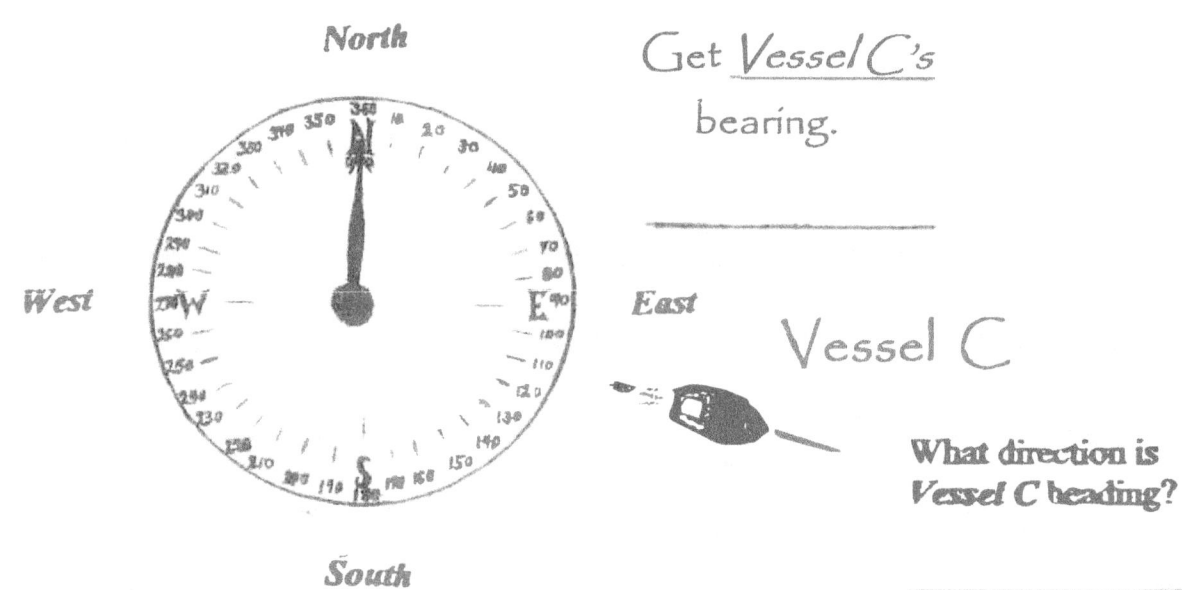

Get _Vessel C's_ bearing.

_____

Vessel C

What direction is Vessel C heading?

_____

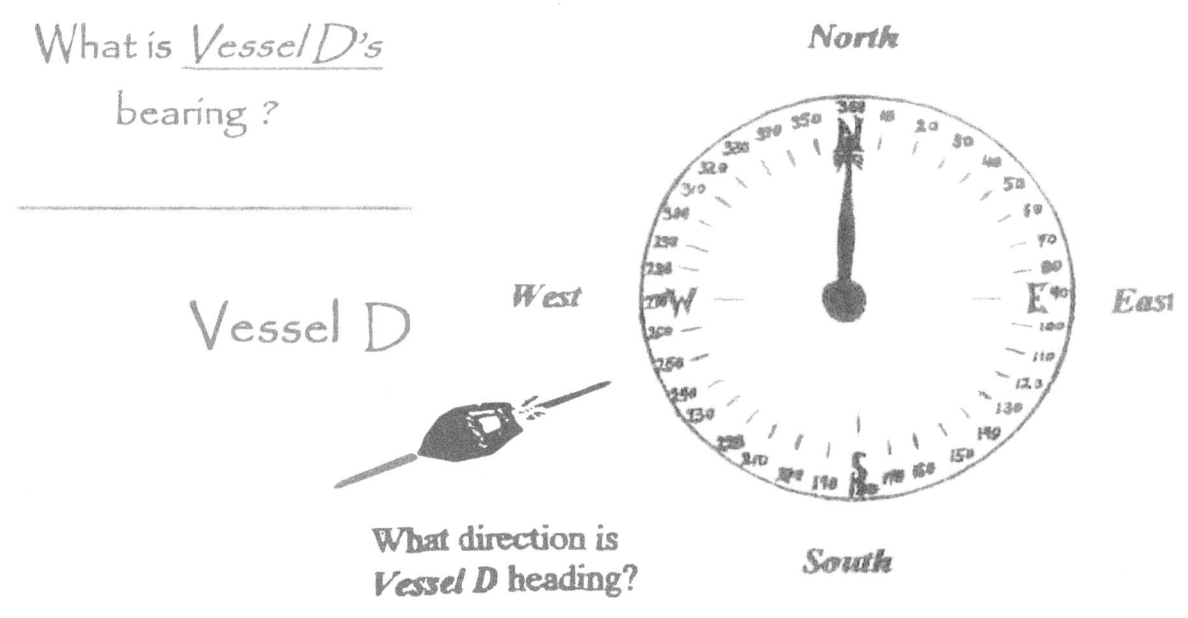

What is _Vessel D's_ bearing ?

_____

Vessel D

What direction is Vessel D heading?

_____

Compass Test
North, East, South, & West
"Headings" and _"Bearings"_
"Answers"

**What is the heading
on _Vessel A_ ?**

310

What is the bearing

on _Vessel A_ ?

310

Vessel A

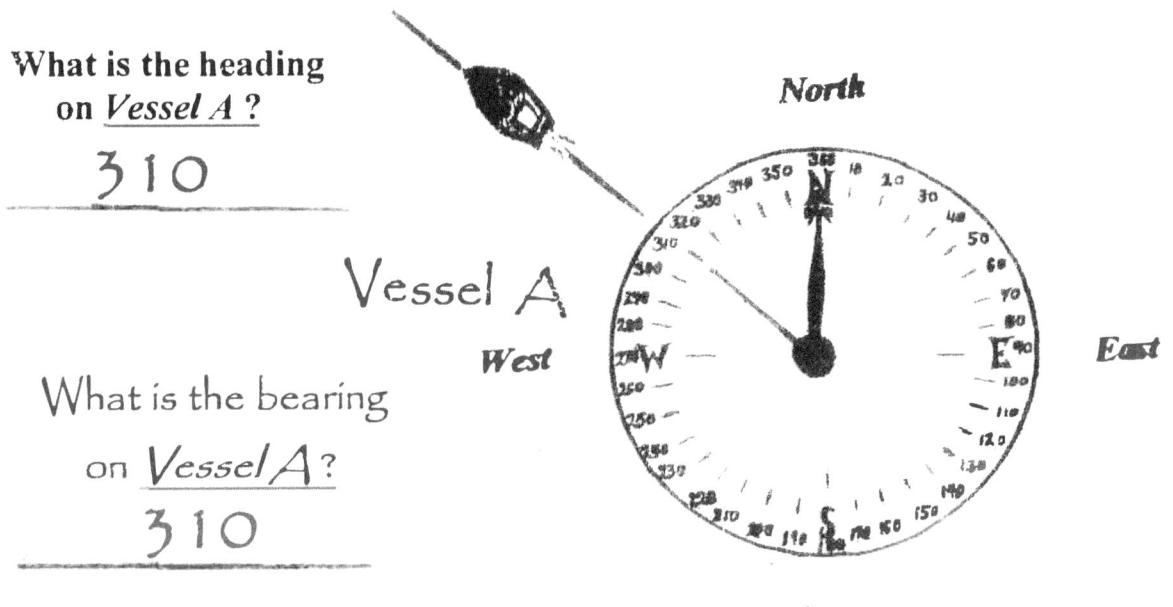

Vessel B

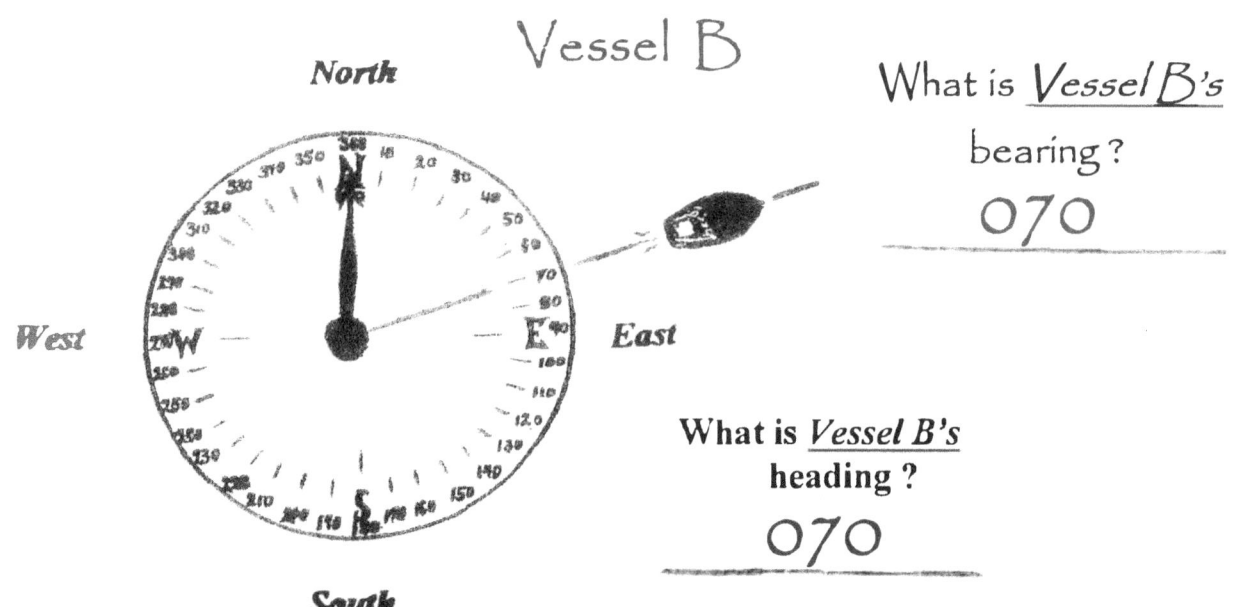

What is _Vessel B's_
bearing ?

070

**What is _Vessel B's_
heading ?**

070

Compass Test

North, East, South, & West
"Headings" and "Bearings"
"Answers"

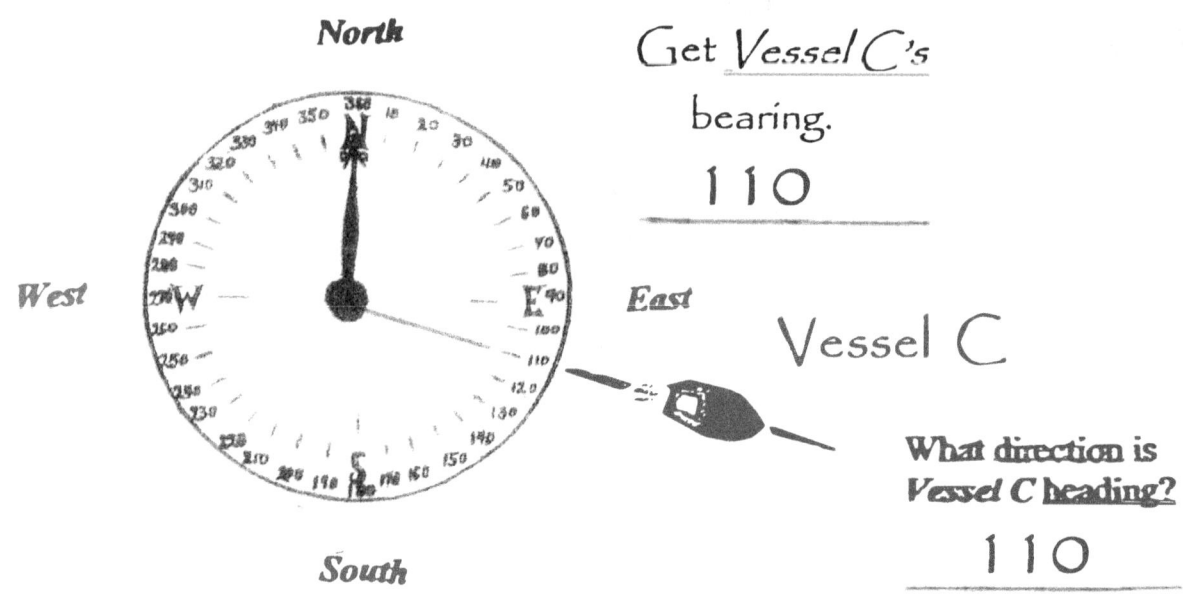

North

Get *Vessel C's* bearing.

110

West

East

Vessel C

South

What direction is *Vessel C* heading?

110

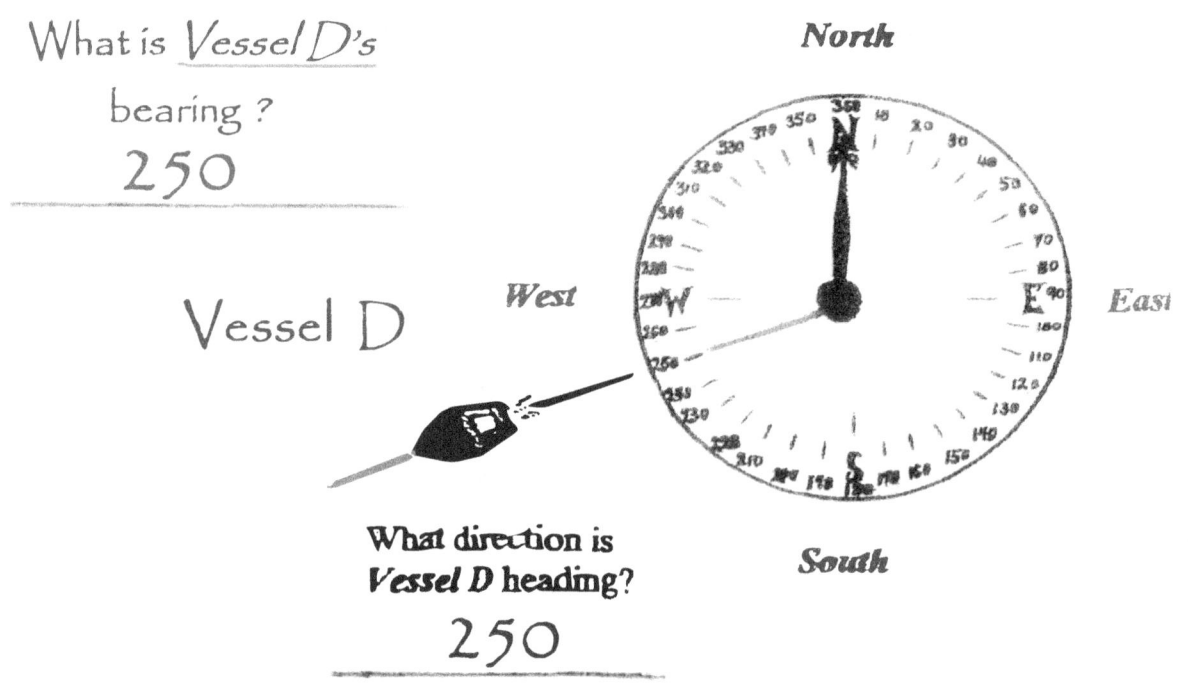

What is *Vessel D's* bearing?

250

Vessel D

West

North

East

South

What direction is *Vessel D* heading?

250

T V M D C

T = True: means or references our vessel's True course as per a chart.

V = Variation: means or references an inaccuracy caused by or within earthly curvatures and earth's varying composition.

M = Magnetic: means or references a vessel's magnetic compass reading.

D = Deviation: means or references an inaccuracy caused by our vessel's construction, magnetic field, or vessel equipment, etc.

C = Compass : means or references our vessel's standard compass course.

Compass Correction Discussion: Captain

T V M D C

The above five categories are the standard for compass correction. The following are memory aids: Our United States Naval Academy teaches;

Truly Valiant Marines Don't Cry At Weddings. At Weddings representing *"Add West"*. Our United States Navy sanctioned the use of; Timely Vessels Make Distance Count At War. At War representing *"Add West"*. The difference ; ERROR, between any two compass directions can be named as follows; *"Compass least, error east"* and *"Compass best, error west"*.

221   009E   002W

T   V   M   D   C

221   009E 212   002W 214

In the above situation our chart gave us two things; the true course between our two pointes as 221 degrees and our variation as East 009 degrees. We knew our deviation as West 002 degrees. We needed to know our magnetic course and our compass course.

Now the other way around.

When we only know our underway heading 127 degrees per standard compass, our chart gave our variation as West 004 degrees. We knew our deviation as East 016 degrees. We use this different memory aid: Can Dead Men Vote Twice At Elections. "At Elections" representing "Add East".

127 016E   004W

C   D   M   V   T

127 016E   143   004W 139

Dutton's Nautical Navigation: 15th Edition,
Rules for Applying Compass Error, #710, Pages 76-83

Compass Correction Discussion Worksheet

T   221 True by Chart
+ W - 009 E Variation
    M = 212 Magnetic Compass

M   212 Magnetic Compass
+ W   002 W Deviation
    C = 214 Standard Compass

C   127 Standard Compass
+ E   016 E Deviation
    M = 143 Magnetic Compass

M   143 Magnetic Compass
+ E - 004 W Variation
    T = 139 True by Chart

The following test vessels; <u>Vessel E</u> and <u>Vessel F</u> are no longer equally aligned with our vessel's position by bearing and heading.

Both <u>Vessel E</u> and <u>Vessel F</u> represent the greatest majority of traffic situations you will find in water traffic.

If you do not have any navigation tools or plotting apparatus to work with, simply use the best common sense evaluation you can muster.

<u>Heading evaluations without any navigation tools</u>

Place an imaginary compass over the vessel of interest in question to determine that vessel's heading.

<u>Bearing evaluations without any navigation tools.</u>

Use any straight edge or imagine a straight edge to align the structural center of the vessel of interest with our compass rose center.

Compass Test
North, East, South, & West
"Bearing"

Vessel E

What is *Vessel E*'s bearing?

_____

Date: _____
Time of Observation: _____
24 Hour Time

North

West

E East

South

North

West W

E East

South

Vessel F

What is *Vessel F*'s bearing?

_____

Pre-study Test Grade: _____
After-study Test Grade: _____

Compass Test
North, East, South, & West
"Heading"

Vessel E

What is the heading
of _Vessel E_?

_____

Date: _____
Time of Observation: _____
24 Hour Time

North

West

East

South

North

West

East

South

Vessel F

What is the heading
of _Vessel F_?

_____

Pre-study Test Grade: _____
After-study Test Grade: _____

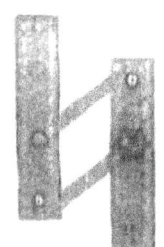

This tool is called a

ParaLock Plotter

by Weems & Plath

We will use a ParaLock Plotter or a conventional parallel rule for this Heading sector with Vessels E and F. Simply unlock the lock knob. Adjust the two sides until you can align the vessel's heading on one side and the other side with the center of the compass rose.

Radar Collision Avoidance is introduced in Lesson #7. In the real world your radar station will help you answer such questions as Heading, Bearing, and Range.

When we transfer a desired course to a compass rose we walk the plotter or parallel rule there. We also start at the Compass Rose and walk our way to a position of interest. We will study

Walking our Plotter, Walking our Dividers, and Walking our Protractor Triangles in greater detail in Lesson #3.

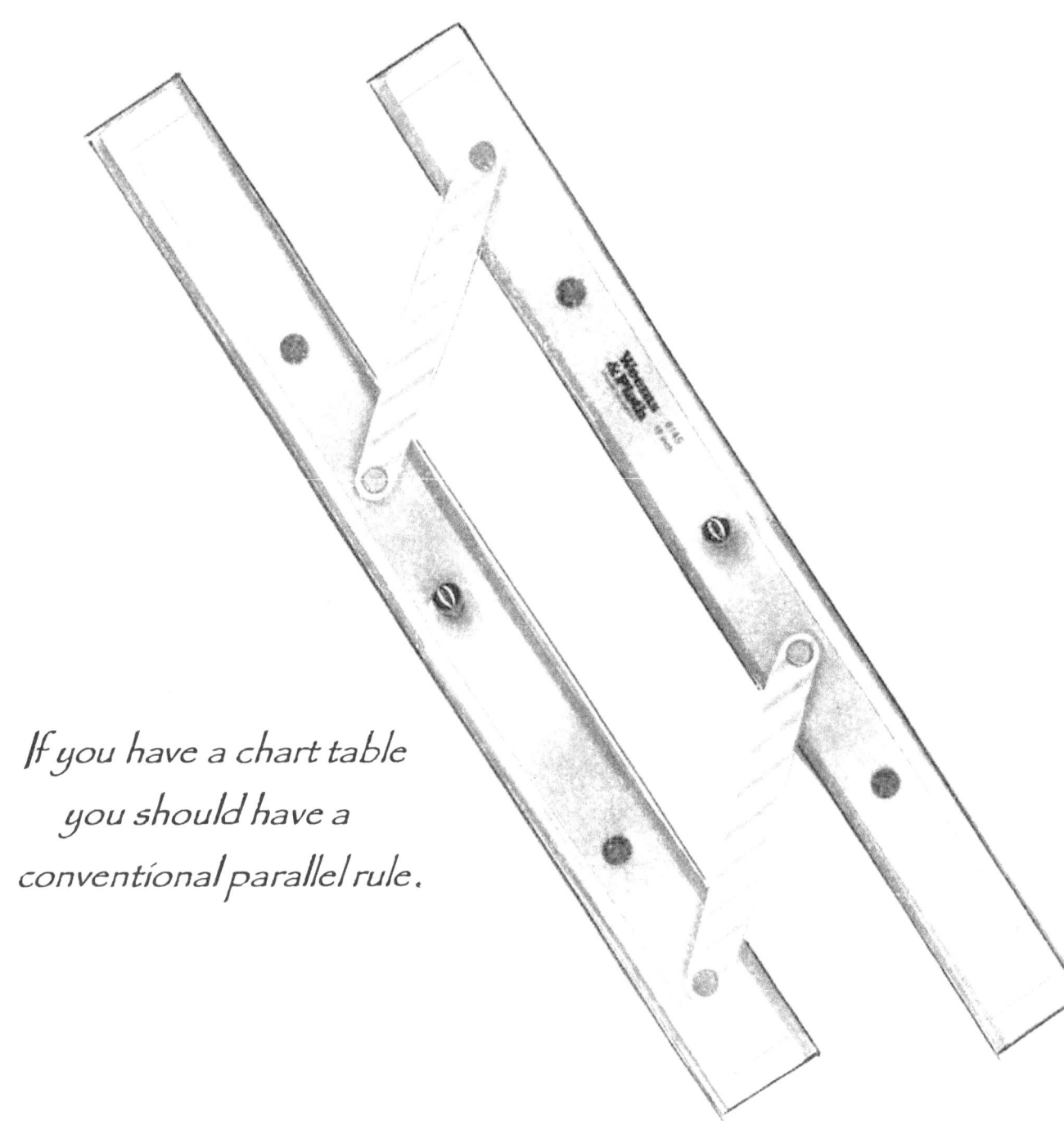

If you have a chart table
you should have a
conventional parallel rule.

"See you in Lesson #3"

Compass Test
North, East, South, & West
"Heading"

"Answers"

Vessel E

What is the heading
of _Vessel E_ ?

020

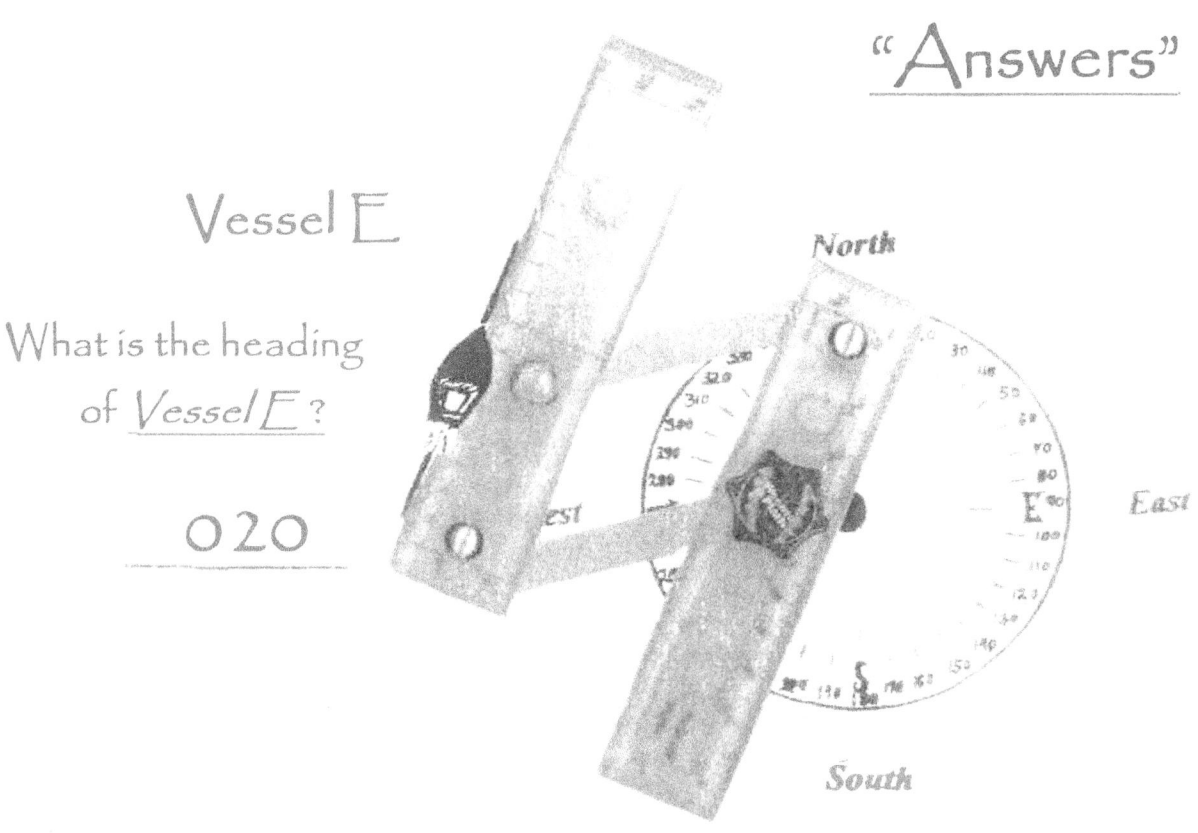

Vessel F

What is the heading
of _Vessel F_ ?

320

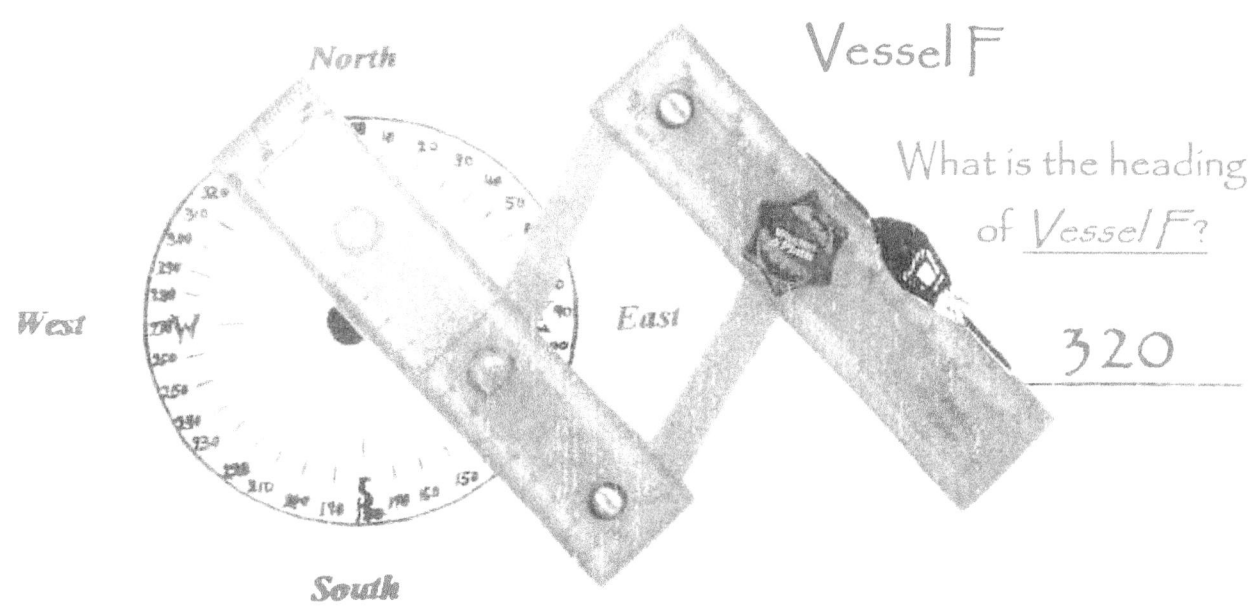

This is a

Weems & Plath 101
Protractor Triangle

#1: Align the long edge of the protractor triangle with the heading of the vessel of interest. This is the base triangle.

#2: Use the appropriate side of the base triangle to slide the traveling triangle towards the compass rose.

#3: The arrows depict the motion of both the base and traveling triangle as they slide their way to the compass rose.

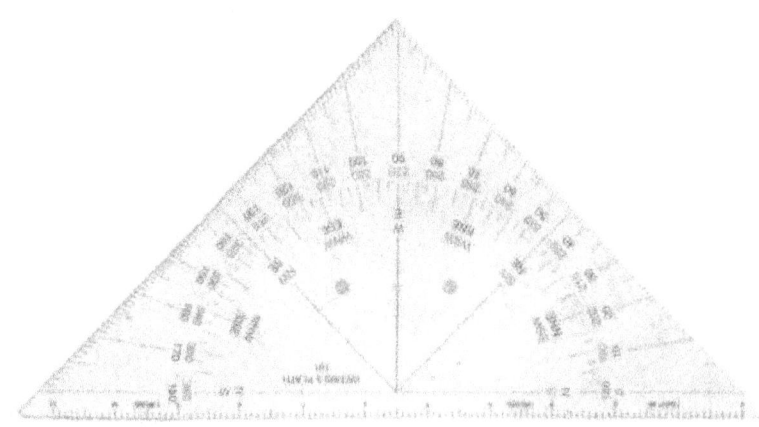

Meridian Course Transfer Procedures
"Weems & Plath 101"
Protractor Triangle

#1: Check to see that the good side; readable side, is up.

#2: Choose a meridian close to your course application. For all purposes in this section a meridian is a longitude.

#3: Put the protractor triangle's 90 degree angle to the right for starters. Then look down to the lower 45 degree angle and memorize the black print indicating 000 - 180 degrees and the red print indicating 180 - 360 degree angles or courses.

#4: Place the central merging point { } on that meridian.

#5: Read the numbers in black print as they represent the angle formed on the upper half of the protractor triangle. The lower half of the same angle will be found in red print just above the black. They are found starting at 000 degrees in black and 180 degrees in red.

#6: Transfer the desired angle to the desired location by sliding the traveling protractor triangle across the flat edge of the base triangle. Then slide the base triangle against the traveling triangle forward to forward until you reach the desired location. If you need to move in another direction use another appropriate side of the base triangle.

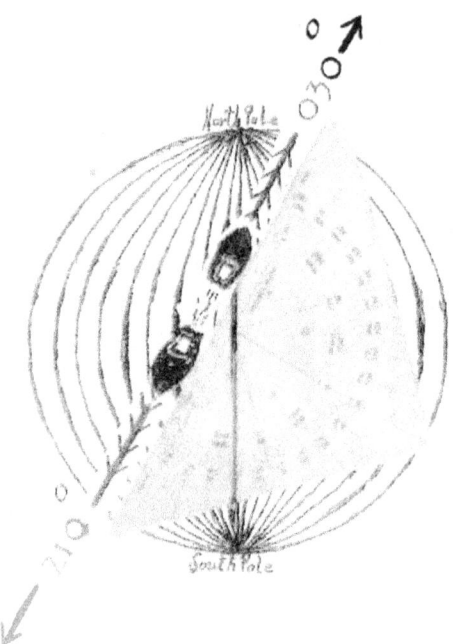

Theoretically you could use any meridian on earth.
Realistically you want to use the closest meridian to your
location of interest. Everything remains relative to your chart.

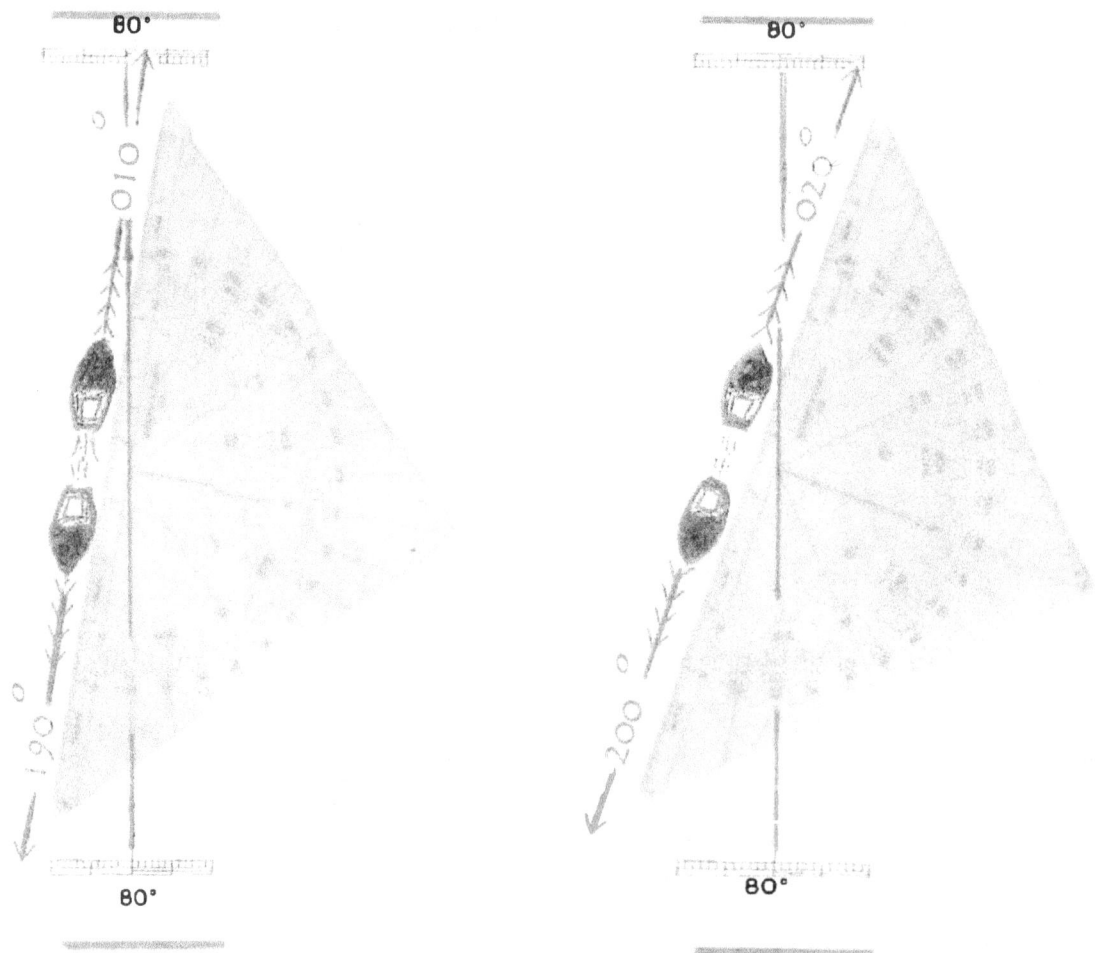

Compass Test
North, East, South, & West
"Bearing"
"Answers"

Vessel E

What is _Vessel E's_ bearing?

290

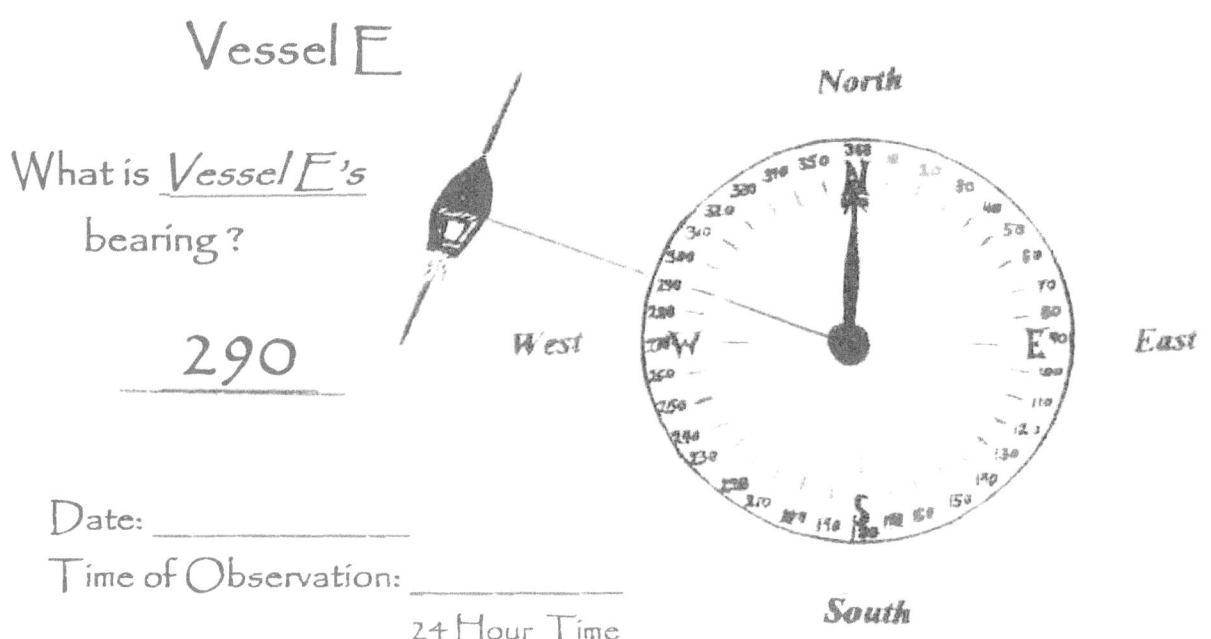

Date: _____
Time of Observation: _____
24 Hour Time

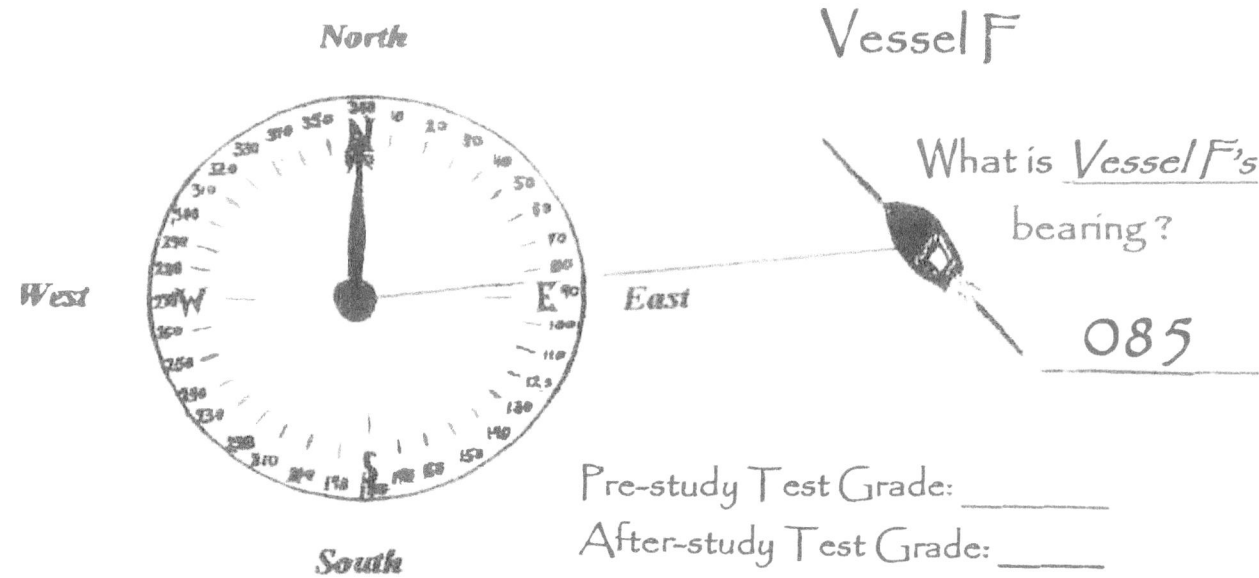

Vessel F

What is _Vessel F's_ bearing?

085

Pre-study Test Grade: _____
After-study Test Grade: _____

There are many ways to use this

Protractor Triangle.

We are going to study just one more.

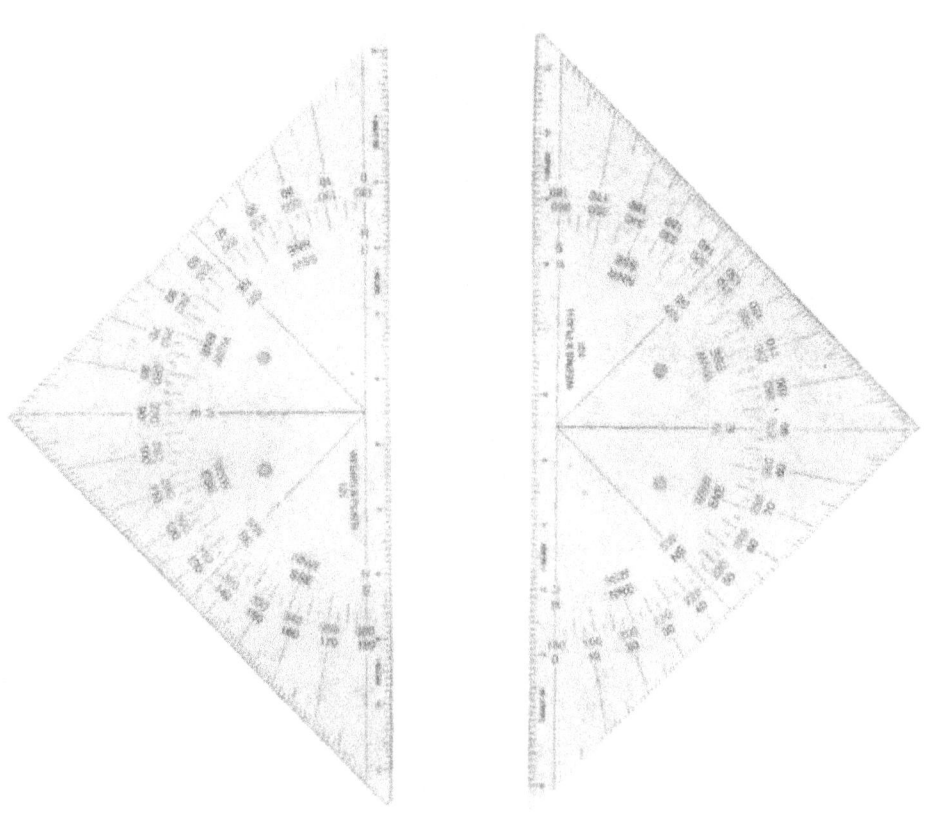

Compass Test
North, East, South, & West
"Heading"
"Answers"

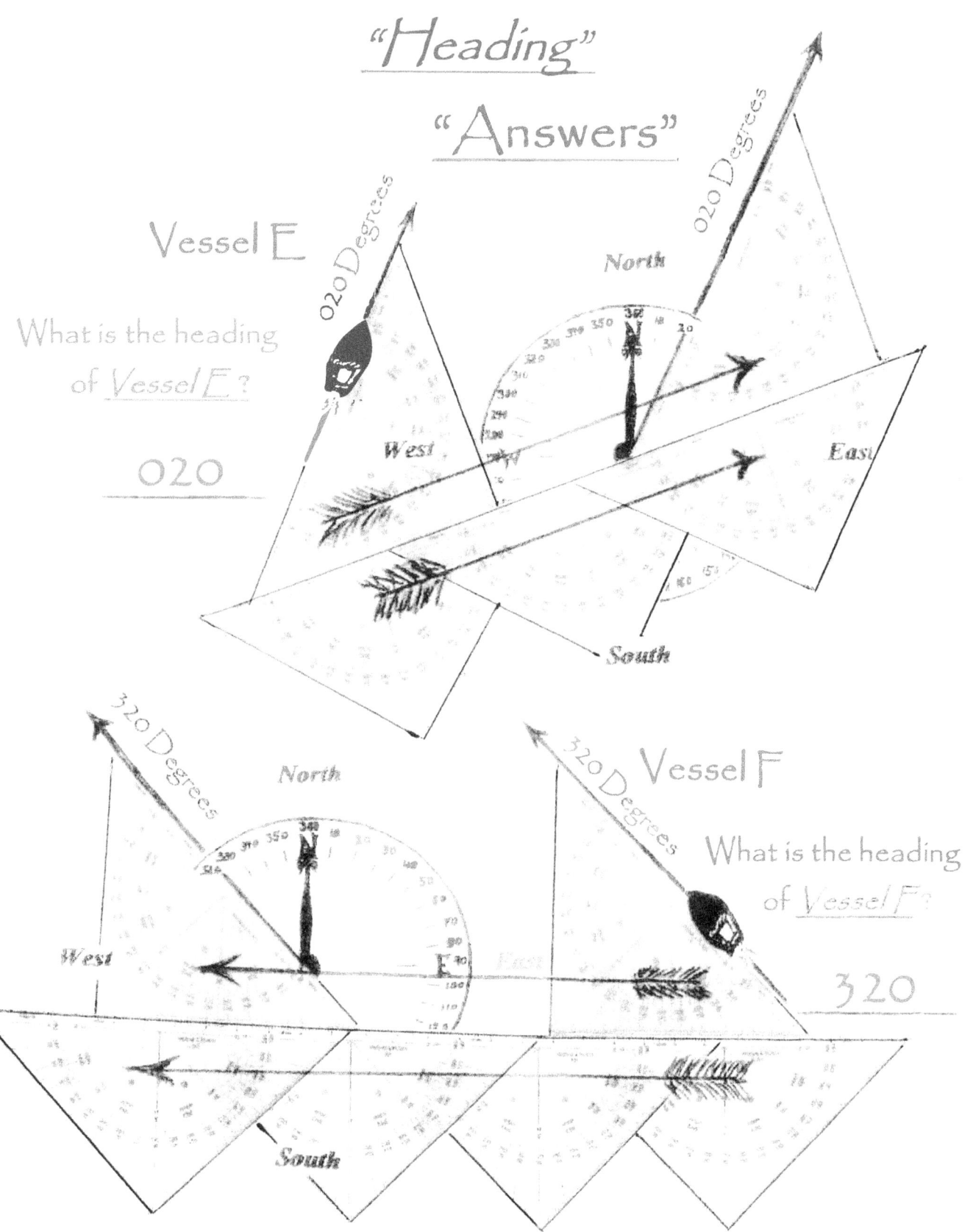

Vessel E

What is the heading of *Vessel E*?

020

Vessel F

What is the heading of *Vessel F*?

320

Compass Vocabulary Test

Grades: 6 – 12 – Captain

Direction: _____

_____

_____

_____

_____

Bearing: _____

_____

_____

_____

_____

Heading: _____

_____

_____

_____

_____

Course: _____

_____

_____

_____

_____

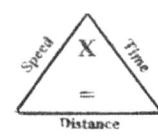 

Compass Correction Vocabulary Test:

Grades: 6 – 12- Captain

T V M D C

T:_____

_____

_____.

V:_____

_____

_____.

M:_____

_____

_____.

D:_____

_____

_____.

C:_____

_____

_____.

This is a Compass Correction question. This is not a Navigation Triangle question.
This is not a Compass Vocabulary question as is the preceding page.
This is a Compass Correction question.

Compass Vocabulary Test Answers
Grades 6-12- Captain

Direction: __Any of the 360 degrees of the compass.__
__North, East, South, and West are the cardinal directions.__

Bearing: __The direction; compass reading, taken from you;__
your vessel, towards an object of interest at a particular moment
expressed in degrees 000 to 360 degrees.

Heading: __The direction in which a vessel points or heads at__
any instant, expressed in angular units, 000-degrees clockwise
through 360-degrees, from a reference point.

Course: __Your intended direction, while your heading is the__
actual direction that you are steering at any given instant.

Compass Correction Vocabulary Test:
Grades 6-12- Captain

T V M D C

T: T = True. T means or references our vessel's True course as per a chart.

V: V = Variation. V means or references an inaccuracy caused by or within earthly curvatures and earth's varying composition.

M: M = Magnetic. M means or references a vessel's magnetic compass reading.

D: D = Deviation. Deviation means or references an inaccuracy caused by our vessel's construction, magnetic field, or vessel equipment, etc.

C: C = Compass. C means or references our vessel's standard compass course.

This is a Compass Correction question. This is not a Navigation Triangle question.
This is not a Compass question as is the preceding page.
This is a Compass Correction question.

Today a vessel's compass has no hands.
Today the entire compass rose turns while finding magnetic north.

All 360 degrees; north, east, south, and west, turn together in a liquid.
There is a notch just outside the turning compass rose that indicates the compass direction of your vessel.
That notch follows the vessel's keel towards the vessel's bow.
The liquid filled compass became standard in 1906.

Magnetic compass technology has been with us for about 1,000 years.

Bowditch, Pub. No. 9, The American Practical Navigator, Page #2, Compass, 2002 Bicentennial Edition

Time Zones

Our last focus in this
Great Circle Study and Division will be on
Time Zones.

We simply divided the full circle; a 360 degree day as Earth
spins by the 24-hours. We first studied the passage of time and
space wherein we clocked that day to be 24 hours and
numerically placed that 360 degree circle around earth starting
in Greenwich, England. It actually represents two days;
one day and one night, at the same time.

$360°$

Capt. Greg Musk
of Maine taught
me this one.

15 degrees = 15 $°$
360 degrees = 360 $°$

$$\frac{360°\ Earth}{\{ \bigcirc \times 2 \}} = 15°\ \text{per Time Zone}$$

24 Hours

In every day life there is but one place at a time holding sunrise.
Yet, this 15 degree an hour passage on earth's axis
will help you tell the time by observation.

1200 Noon

Time Zones

This 15 degrees per Time Zone is about the same as saying
the sun appears to travel across our sky at 15 degrees an hour.
Herein, to tell the time by the sun calculate to or from high noon.

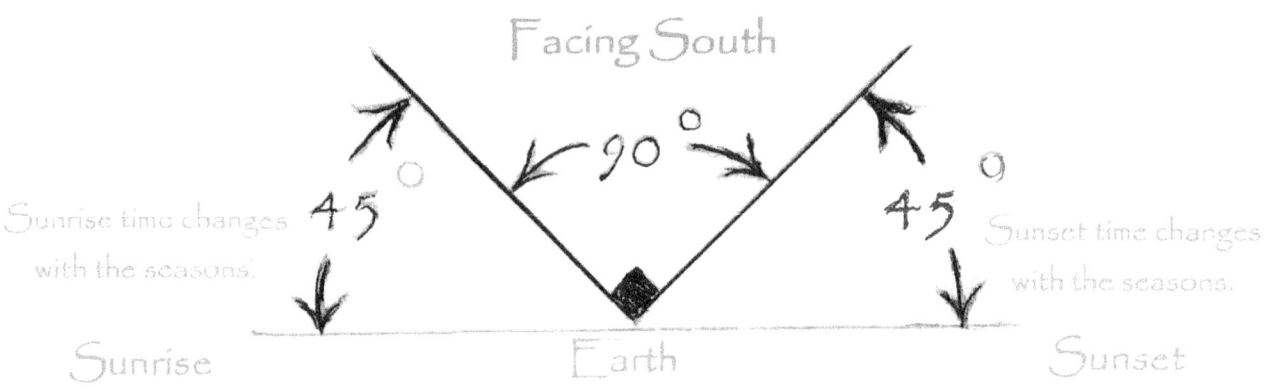

That Time Zone diagram still looks
like a baseball diamond
to me. See you in
Lesson #3.

Do NOT look into the SUN

Time Zones are an advanced navigation study.
Looking at both sides of our earth; both day and night,
at the same time is not an easy task.

Thus, we will now change our focus to an
underway Range and Bearing observation
as seen from a vessel's windows.

Now remember, if you save one life by noticing
this upcoming Range and Bearing observation
you become a real hero and
most likely a real captain.

The Big Picture

Time Zones

{Time Zones}

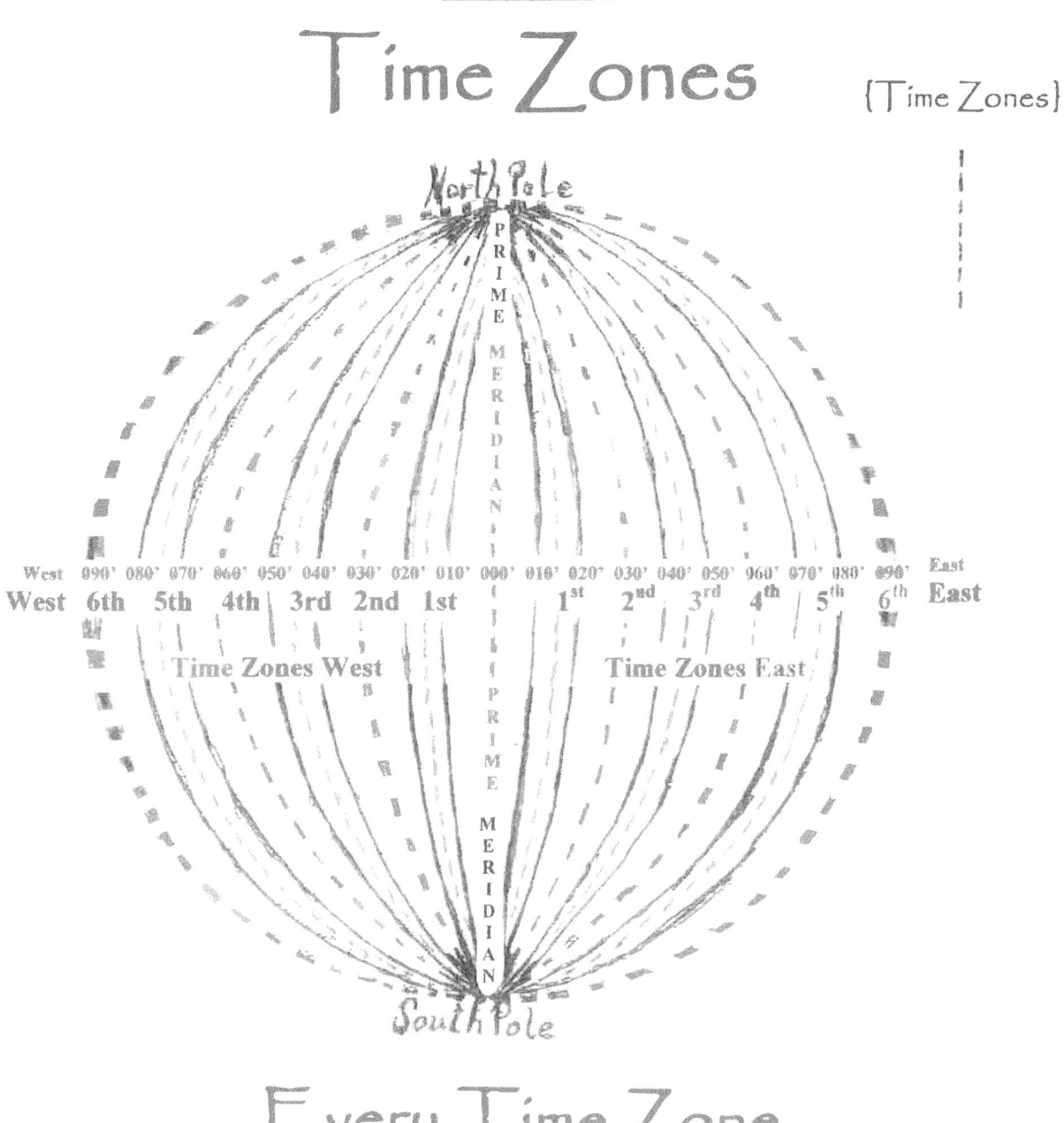

Every Time Zone
has
15-Degrees of Longitude

The Big Picture
always includes far away.

Therefore our focus of interest will
now be Range and Bearing.
Other vessel traffic is our
focus of interest at
this moment.

Range: refers to distance.

Bearing: refers to direction.

While underway;
looking out any window,
if you see the same vessel
never seeming to move but
getting closer you are
experiencing

a

Risk of Collision.

Risk of Collision

Range and Bearing

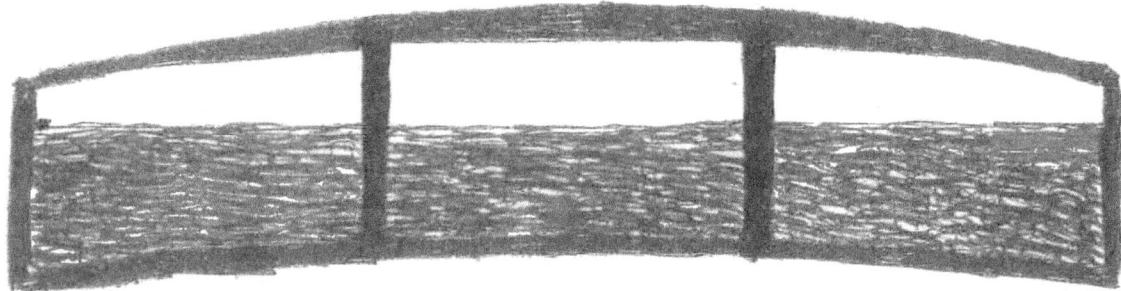

Notice that spec in the window ?

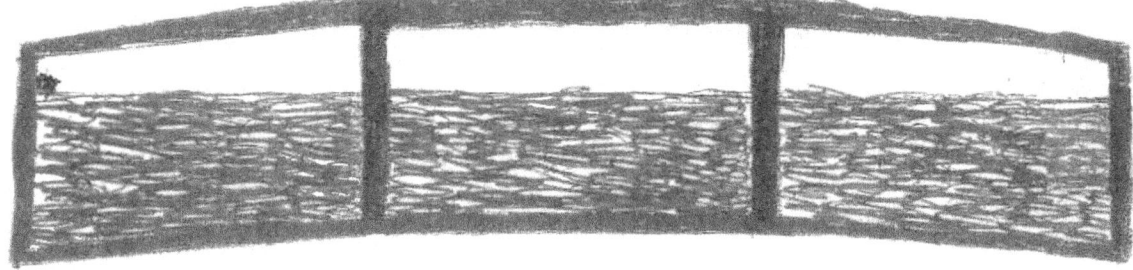

That spec is a vessel on your port.

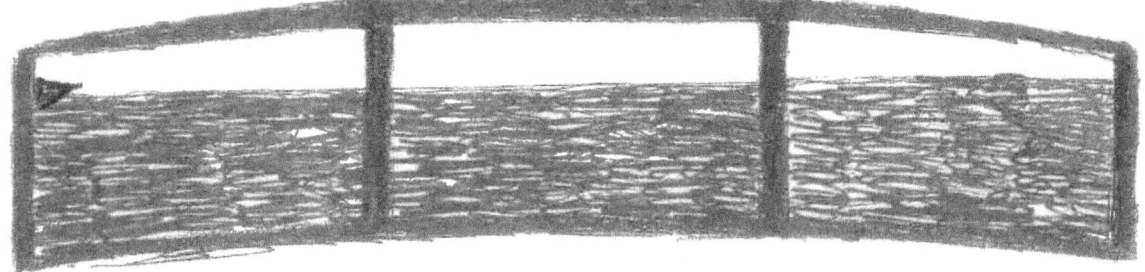

That vessel is on a collision course with you.
That vessel on your port is the burdened vessel and must alter her course.
Your vessel has the Right of Way. However, make radio contact with that vessel to be sure.

Incoming traffic from any direction seen out any window;

when the bearing remains the same and the range

decreases you know that a dangerous

Risk of Collision Exists.

Range & Bearing

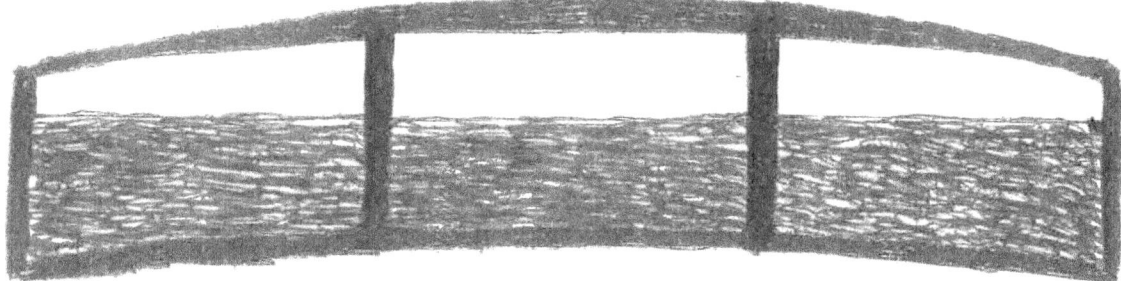

Notice that spec in the window ?

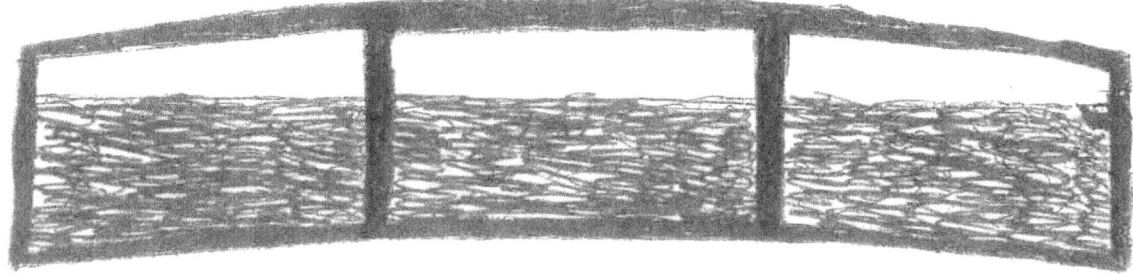

That spec is a vessel on your starboard.

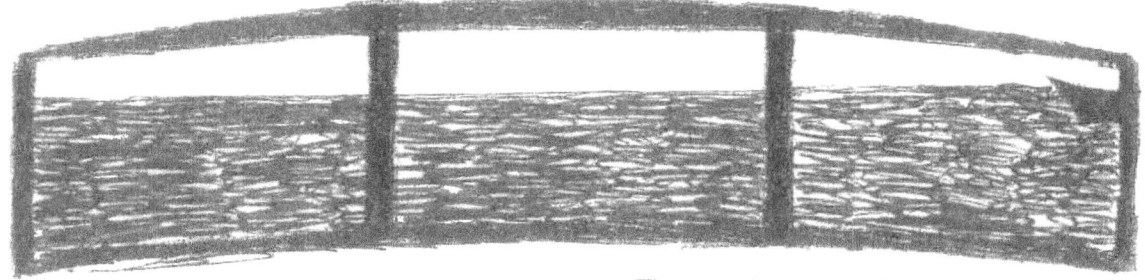

That vessel in on a collision course with you.
That vessel on your starboard has the Right of Way.

If that spec stays in the same spot in the same window;
any window, and just gets larger. You are on a collision course.

"If the Bearing stays the same and the Range decrease, you are on a collision course."

At night all you would see are lights that get brighter
and never seem to move.

Range: *Range refers to distance.*

Bearing: *Bearing refers to direction.*

It is really a small thing when you look at it.

Nautical Checklist

Grades K-6

Put your life preserver on as you
leave your car. ☑

Never run to the dock. ☑
Remember
Never run on the dock. ☑

Never run on the boat. ☑

Try to say "May I ?" and
"Thank You." ☑

Listen to your parents and
listen to the captain. ☑

OK! Everybody else, same as above.
Keep an eye on your K-6 crew.
Know where your lifejacket is.
Do not use your lifejacket
for a seat cushion.

Navigation Triangle
Advanced Navigation

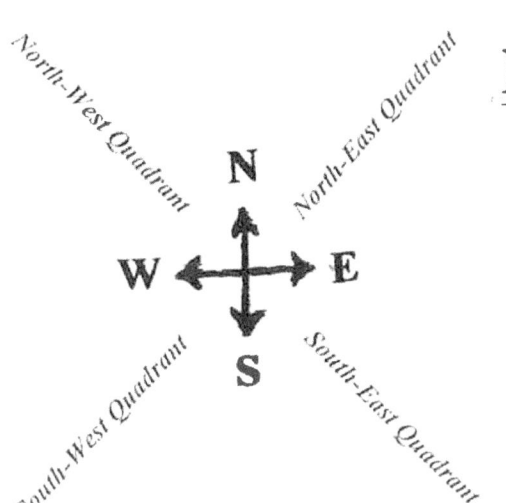

L1 = Departure Latitude

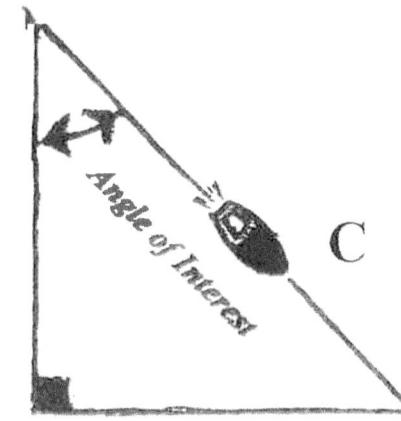

Angle of Interest

B is a Longitude
B is measured as
D'Lat
1-Min of Lat = 1-NM
NM = Nautical Mile

B

C

C is our Course-line
C is measured in NMs

NMs = Nautical Miles

L2 = Arrival Latitude

A

A is a Latitude
A is measured as D'Long in NMs

{ D'Lat } means:
The difference between latitudes.
Found by subtracting one latitude
from the other.

{ D'Long } means:
The difference between longitudes.
Found by subtracting one longitude
from the other.

Navigation Triangle

Advanced Navigation
and
Trigonometry

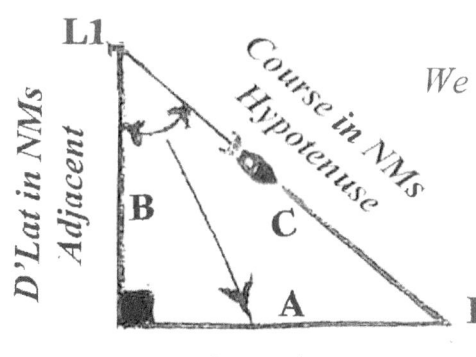

L1

D'Lat in NMs
Adjacent

Course in NMs
Hypotenuse

B

C

A

L2

We are always interested
in one angle and two sides
when working a
Navigation Triangle.

Opposite
D'Long in NMs

Adjacent, Opposite, and Hypotenuse will be studied on following pages

We use a Common Denominator of Nautical Miles
1-Minute of Latitude = 1-Nautical Mile

All Great Circle Voyages approach a
090 leg or a 270 degree leg.
That 090 or 270 degree leg is called the Vertex.

The Vertex is either the highest latitude
or the lowest latitude reached in a Great Circle Voyage.
We plot a Great Circle ocean crossing in Lesson #5.

Navigation Triangle
Advanced Navigation

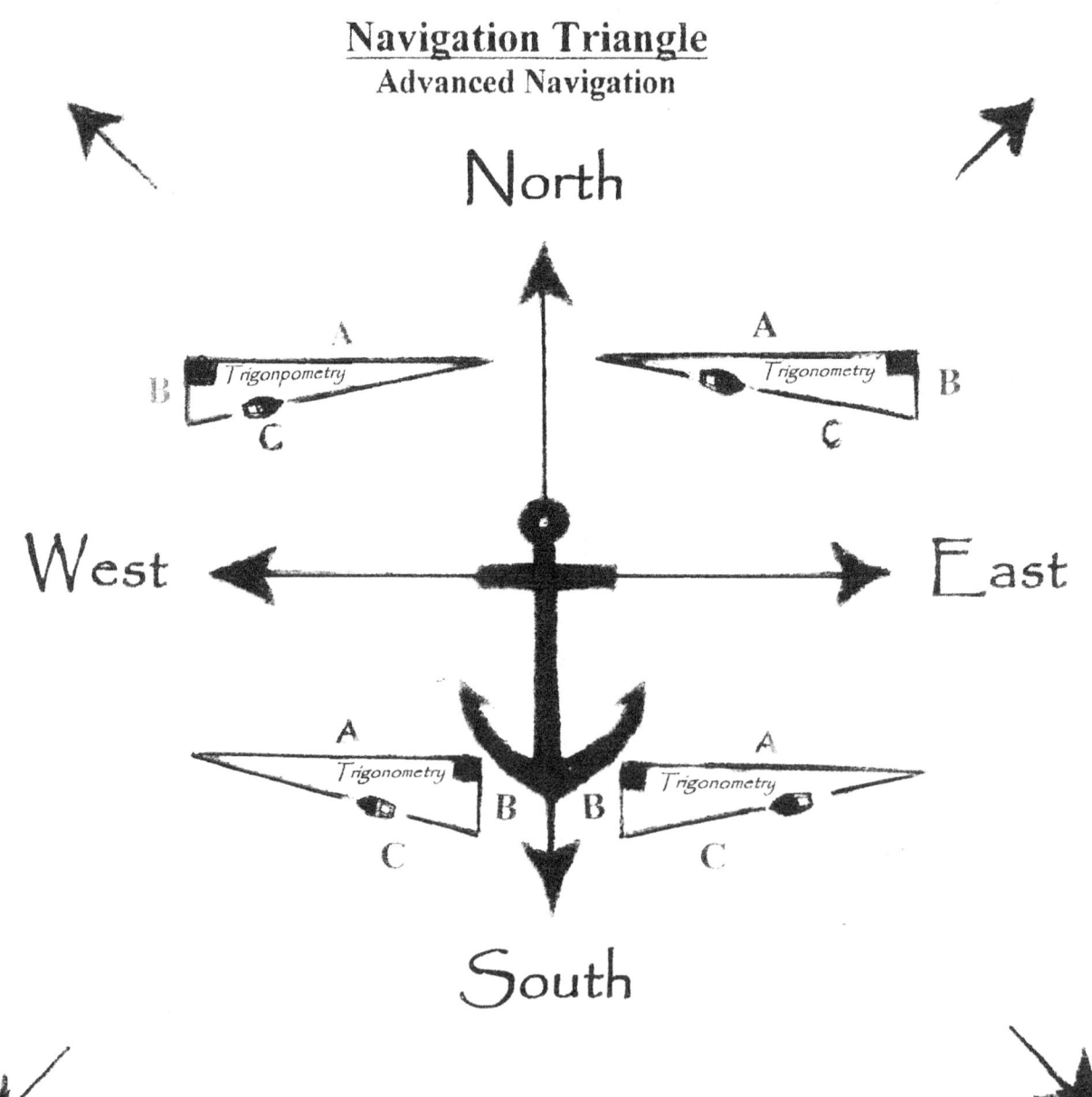

Navigation Triangle
Advanced Navigation and Trigonometry

Captain Soh Cah Toa: Adopted K.I.S.S. format, "Keep It Simple Sailor."
Captain Soh Cah Toa is actually code to the Nautical Triangle.

We are always interested in
1-Angle {Either L1 or L2} and **2-Sides** of the Navigation Triangle.

<u>**Soh**</u> means: <u>**S**</u>ine Angle = <u>**O**</u>pposite side divided by the <u>**H**</u>ypotenuse.
<u>**Cah**</u> means: <u>**C**</u>osine Angle = <u>**A**</u>djacent side divided by the <u>**H**</u>ypotenuse.
<u>**Toa**</u> means: <u>**T**</u>angent Angle = <u>**O**</u>pposite side divided by the <u>**A**</u>djacent side.

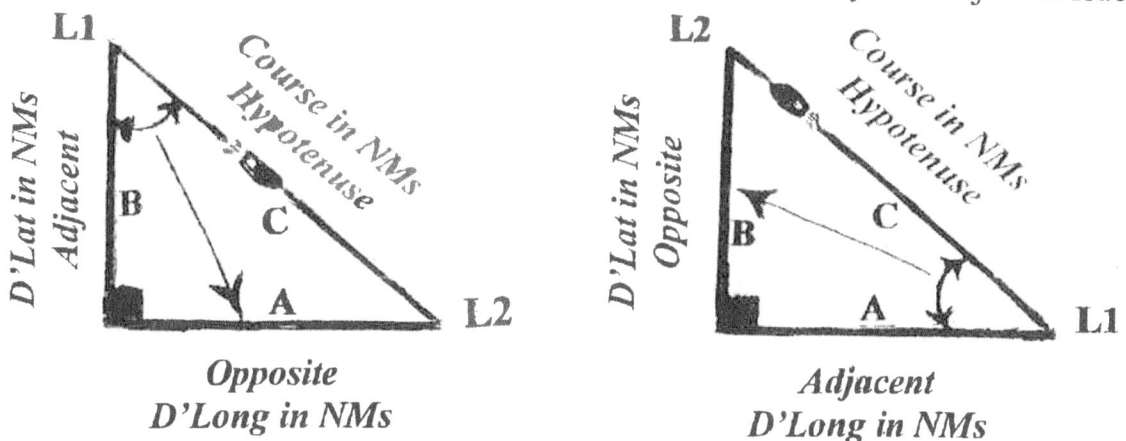

NMs = Nautical Miles

Adjacent: The side of our angle that connects to the <u>90-Degree Angle</u> where Latitude and longitude meet.

Hypotenuse: Always the Course-line {on charts} and a <u>Line of Sight Observation to a Celestial Observation.</u>

Opposite: The 3rd side of our Navigation Triangle that closes our angle of focus into a triangle. The side connected to the <u>90-Degree Angle</u> and he NOT part of the angle of focus.

In the navigation Triangle we use a Common Denominator of Nautical Miles where 1-Minute Latitude = 1-Nauticle Mile

The formula for these parts of D'Long is: **p = parts**
p = D'Long in NMs = Cos Mid-Lat x D'Long in Mins = NMs
Page #581: American Practical Navigator, Bowditch, Volume #2

$$Mid\text{-}Lat = \frac{L1 + L2}{2} = lm$$

Advanced Navigation

Compass Notation

The Navigation Triangle by *Captain Soh Cah Toa*

*Compass Notation takes our answer from a **90-Degree Relativity** to a **360-Degree Compass** relativity. **Common Sense** will form your choice of **Quadrants** in which to apply your answer. This common sense decision is derived from knowing what direction your **L1 and L2 locations** lead.*

First label your answer; indicate a N or S direction.
Complete that label; indicate the E or W quadrant you are heading within.
Compass Notation places our answer relative to our **360-Degree Compass.**

Designated Answer 067-Degrees

*I chose the answer of 067-Degrees because I am 67-yrs old, no other reason. Let us say **Common Sense** and **Deductive Reasoning** tell us that we are headed towards and with-in the **South-West Quadrant.***

{Below we will see how an answer relative to 90-Degrees via sin, cos, and tan can be transferred to our 360-Degree Compass for navigation purposes or for USCG exam purposes}.

NW Quadrant.
Compass Notation {N Answer W}

North-West Quadrant
N = 360-Degrees
Minus "Your Answer"
*If you were headed in a **NW'rly** direction you would then have a **Compass Rose** answer **Relative to 360-Degrees***

North

NE Quadrant
Compass Notation {N Answer E}

North-West Quadrant
N = 000-Degrees
Plus "Your Answer"
*If you were headed in a **NE'rly** direction you would then have a **Compass Rose** answer **Relative to 360-Degrees***

West ← → **East**

SW Quadrant.
Compass Notation {S Answer W}

South-West Quadrant
S = 180-Degrees
Plus 067-Degrees
247-Degrees Compass Course
*Now you have a **Compass Rose** answer **Relative to 360-Degrees***

SE Quadrant
Compass Notation {S Answer E}

South-East Quadrant
S = 180-Degrees
Minus "Your Answer"
*If you were headed in a **SE'rly** direction you would then have a **Compass Course** answer **Relative to 360-Degrees***

South

73

Compass Notation

Advanced Navigation

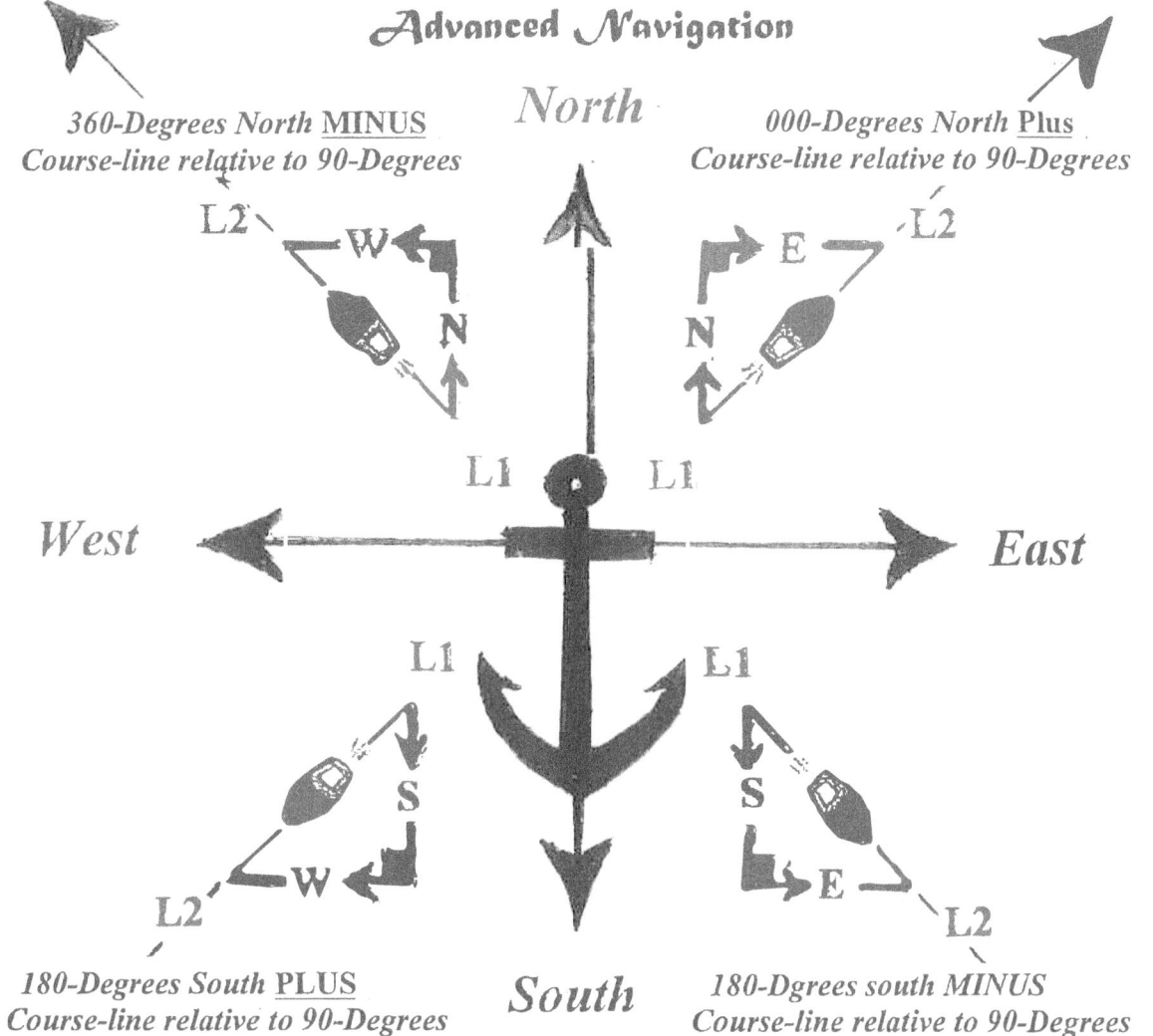

North

360-Degrees North <u>MINUS</u>
Course-line relative to 90-Degrees

000-Degrees North <u>Plus</u>
Course-line relative to 90-Degrees

West

East

180-Degrees South <u>PLUS</u>
Course-line relative to 90-Degrees

South

180-Dgrees south MINUS
Course-line relative to 90-Degrees

Common Sense

As Captain your first responsibility is
The Safety of Life at Sea.
As a Captain that responsibility is also to follow the
United States Coast Guard Navigation Rules; International and Inland.
Then comes common sense.
As the captain you are obligated to know where the
north, east, south, and west are at all times, in all conditions.
At that point, you already know what quadrant you are sailing towards;
being either the NE, NW, SE, or SW quadrant. There-in you use common
sense to apply your **compass notation** *from the 90-Degree relativity of the*
Navigation Triangle to the 360-Degree relativity of our
Navigation Compass.

Two Boats School – Two Circles of Time

Student's Name: _____

Date: _____

1, 2, 3, & A, B, C

Daily Test

Grade in School: _____

Previous Test score: _____

Today's Test Score: _____

Navigation Triangle

A = _____ **on Charts**

B = _____ **on Charts**

C = _____ **on Charts**

Line of sight observation — **in Celestial Navigation**

The shortest distance between any 2-Points is a _____.

The shortest distance between any 3-points is a _____.

Lesson #1 *Home Schooling* Test

How many degrees are there in the following ?

Student's Name _____
Date: _____
Grade in School: _____

Lesson #1 Test-Grade: _____

#1: North = _____

#2: East = _____

#3: South = _____

#4: West = _____

#5: North-East = _____   {[000-degrees + 90-degrees] / 2} = 45-degrees

#6: South-East = _____   {[180-degrees + 90-degrees] / 2} = 135-degrees

#7: South-West = _____   {[180-degrees + 270-degrees] / 2} = 225-degrees

#8: North-West = _____   {[360-degrees + 270-degrees] / 2} = 315-degrees

#9: Write out Pythagorean Theorem . _____

#10: How many degrees are there in a Right Triangle ? _____

#11: Every Right Triangle has a certain kind of angle in it.
That particular angle is always a _____ .

#12: There are how many degrees in a circle ? _____

#13: 1-Nautical Mile is = to _____ of Latitude ?

#14: To take a celestial observation.
You, measure the celestial objects _____ above the _____ .

#15: Numerically Latitudes run _____ & _____ .

#16: Numerically Longitudes run _____ & _____ .

#17: There are how many degrees of Lat. ? _____ -North & _____ -South.

#18: There are how many degrees of Long ? _____ -East & _____ -West.

#19: Vertical lines on a chart stretching N. & S. {like this ⬤} are: _____

#20: Horizontal lines on a chart stretching E. & W. {like this ⬤} are: _____

Two Boats School – Two Circles of Time

1, 2, 3, & A, B, C

Daily Test
ANSWERS

Navigation Triangle

A = _____ *Latitude* _____ **on Charts**

B = _____ *Longitude* _____ **on Charts**

C = _____ *Course Line* _____ **on Charts**

Line of Sight Observation **in Celestial Navigation**

The shortest distance between any 2-Points is a ___ *Straight Line* ___.

"We steer in a straight line to save time and fuel."

The shortest distance between any 3-points is a ___ *Triangle* ___.

"With 2-known factors in a navigation triangle we can find much more than just a third." "Remember: Common Sense saves time & lives."

Lesson #1 *Home Schooling* Test Answers

How many degrees are there in the following ?

Student's Name: _____
Date: _____
Grade in School: _____

Lesson #1 Test-Grade: _____

#1: North = *360-degees or 000-degrees*

#2: East = *90-degrees*

#3: South = *180-degrees*

#4: West = *270-degrees*

#5: North-East = *45-degrees*   {[000-degrees + 90-degrees] / 2} = 45-degrees

#6: South-East = *135-degrees*   {[180-degrees + 90-degrees] / 2} = 135-degrees

#7: South-West = *225-degrees*   {[180-degrees + 270-degrees] / 2} = 225-degrees

#8: North-West = *315-degrees*   {[360-degrees + 270-degrees] / 2} = 315-degrees

#9: Write out Pythagorean Theorem. $\{A^2 + B^2 = C^2\}$

#10: How many degrees are there in a Plane Right Triangle ? *180-degrees*

#11: Every Plane Right Triangle has a certain kind of angle in it.
     That particular angle is always a *{90-degree angle}* .

#12: There are how many degrees in a circle ? *{360-degrees}*

#13: 1-Nautical Mile is = to   *{1-Minute of}*   Latitude.

#14: To take a celestial observation.   *Angle*
     You, measure the celestial objects - *Height* - above the - *Horizon* -

#15: Numerically Latitudes run   - *North* - & - *South* - .

#16: Numerically Longitudes run -*East* - & - *West* - .

#17: There are how many degrees of Lat. ? *{90-degrees South}* & *{90-degrees North}*

#18: There are how many degrees of Long ? *{180-degrees East}* & *{180-degrees West}*

#19: Vertical lines on a chart stretching N. & S. {like this ⬤ } are: *Longitudes* .

#20: Horizontal lines on a chart stretching E. & W. {like this ⬤ } are: *Latitudes.*

Common Sense

Common Sense is independent of technical training and free of intellectual study. Common Sense followed the rules before the rules became a matter of law. Common Sense is good judgment.

Safety

Boating safety and **Life Jackets** are one in the same.
Always have at least one life-jacket per-person.

Life jackets are a single use item. Do not use your life jacket for a seat cushion. That may crush the flotation material that is there to keep you afloat. Once you have been assigned your quarters your first safety move is a **1, 2, 3 basic safety move.**

1st: Know where your life jacket is stored. Then make certain it fits and is in proper condition. Feel it with your hands and try it on. Make certain that the outer material is still fit for rough weather. Take some of the outer fabric in your fingers and rub it together. That life jacket is there to save your life in the worst of situations.

2nd: Know where the first fire extinguisher and first fire station is located outside your door. Read the labels and directions. Imagine what you would do if you found a fire or smelled smoke. Of-course you would inform somebody else and investigate. Always have a back-up person. It takes a team to fight a fire. If there is a fire and you cannot win in 30 seconds go for help.
Nobody wants to find a dead hero, find help.

3rd: Know how to exit your quarters into the passageway and out through a weatherproof bulkhead to your Emergency Muster Station. Know this with smoke so thick you can only see the floor.

Safety

Safety Tip #4: When walking a stairwell; use one hand for your labor and one hand for yourself. Keep one good hand on the handrail. I remember one winter when both my feet slid off the steps beneath me. If I had not had both hands on the handrails I would have been in real trouble. As it was I landed on the handrails as if I was landing on parallel bars in a gym with my elbows out. I slid down the handrails to land feet first on the deck below.

Safety Tip #5: When working aloft; wear a harness. The view is one of the best on a ship. Plus you are almost guaranteed a good breeze. But holding on with one hand while you work with the other gets old fast. Wear a harness when you work aloft & secure yourself with a short fall-line. You will be trained to know all such safety regulations relative to your job description and all safety tips best suited for your specific job description.

Safety Tip #6: When trouble shooting loose cargo in rough weather; do not get between a heavy moving object and the bulkhead. Before the weather gets rough one of your duties will be to lash down everything aboard ship before you run into the rough weather. If by chance a heavy object works loose and is causing havoc you may be asked to temporarily secure that object in rough weather. If you or your team is sent to secure a loose object in rough weather do not to get between that loose object and a bulkhead. Secure a line to the bulkhead and when the object hits the bulkhead secure the object and wedge the underside.
If you sense that something is wrong, tell somebody. If you smell smoke, tell somebody. Tell a superior officer if you notice something that is unsafe and must be entered into a work detail.

Think safety, act as a team, stay safe, and be safe.
Leadership is teamwork.

Leadership is Teamwork

Be determined to learn and you will learn.
Remember: COMMON SENSE SAFTEY is first and
foremost at Sea. It has become a mater of law that books
be carried aboard ships to refresh and preserve standards.

We are now riding the Sea of Math. Before we move on into
deeper waters sharpen your pencils. Our One-room
Schoolhouse Edition makes room for everyone.

Being a captain is an honor and a responsibility.

The most important issue upon a captain today is
the
Safety of Life at Sea.

Simplified Text Follow-up Discussion

First the desired "Quick Answer"

"They are mathematically compatible."

Does anyone know why there are 60-seconds to every minute, 60-minutes to every hour, 12-hours on the face of a clock or watch, 24-hours in every day, 360-Degrees in a Compass,
180-Degrees East Longitude,
180-Degrees West Longitude,
180-Degrees in a Plane Right Triangle,
90-Degrees North Latitude,
90-Degrees South Latitude, and
90-Degrees to a Right Angle ?

Desired "Quick Answer":
"They are mathematically compatible".

We have set sails for an adventure in time on the Sea of Math. Welcome aboard !

Before you become a Captain you will have to serve many hours on deck. You actually have to obtain documented sea-time aboard a vessel from a licensed Captain before you can test for officer's papers. I worked the deck of many fishing vessels to acquire my needed sea-time. Experiencing seamanship on the water is what we call getting your sea-legs. I have stood Bridge Watch on a 700 plus foot ship standing 70 to 80 feet above the face of the Sea. I would report lights as they appeared on the horizon or out of a dense fog. If you were standing Bow Watch as your ship departed your slip into a river and you saw no traffic you may say this:

"Bridge <> Bow"
"Downstream Upstream, All Clear"

Now, if you saw a vessel or many vessels coming downstream you would call the bridge and state your observation like this.

"Bridge <> Bow"
"Downstream Traffic"

On a radio you want to keep your statements short.
This is because the Captain already knows almost everything.
He only wants to hear what he does NOT know.
The less you say, the less time occupied to get the message through, the better job you did.
Do not raise your voice. Speak directly into the speaker. Speak in a calm deliberate voice. Speak and release the speaker button.

Be certain to hold the speaker button down before you speak; wait a second before you speak. Do not be in a race between your finger on the speaker button and your voice.
Let the batteries have a second to do their job.

Next, we will look at how fast a vessel may go.
We refer to our speed over water in **knots.**

1-Knot is 100-Feet per Minute.

US Nautical Mile = 6,000 Ft. & British Nautical Mile = 6,080 Ft.

*Centuries ago a seaman would stand aft-leeward with
a long line with knots every-fathom or so. He might
have a minute glass or use another count.
At a proper sea he would heave the
float and time the knots passed
less the fathoms thrown.
Thus the term
knots.*

*Today, as huge ships approach a dock the Captain
knows that at 1-knot his vessel travels 100-Ft
per minute. At 2-knots his vessel will
travel 200-feet per-minute.*

*Today **100-feet per Knot per Minute** is a Captain's
docking guide, as long as reverse works or tugs are in use.*

*On a boat we would use common sense to approach
the dock at a safe speed, a slow no wake speed.
100-feet a minute is safe lest you need
to maneuver for larger vessels.*

Knots per Hour.
*On the water; **1-Knot = 1-Nautical Mile per-Hour.***

1-Minute of Latitude represents 1-Nautical Mile.
*More on this important nautical distance factor
in the lessons ahead.*

Two Boats School – Two Circles of Time

Student's Name: _____

1, 2, 3, & A, B, C   Date: _____

Daily Test

Grade in School: _____
Previous Test score: _____
Today's Test Score: _____

Navigation Triangle

A = _____ **on Charts**

B = _____ **on Charts**

C = _____ **on Charts**

_____ **in Celestial Navigation**

The shortest distance between any 2-Points is a _____.

The shortest distance between any 3-points is a _____.

85

Remember: press the speaker button on your radio before you speak. *Speak clearly, do not shout, and keep it short.*

*The **Bow** is the front of your vessel and the **Stern** is the back of your vessel.*

***Captain's Math** uses the **Common Denominator** of*
1-Nautical Mile = 1-Minute of Latitude
1-Knot = 1-Nautical mile per Hour
1-Knot = 100-Feet per Minute

Nathaniel Bowditch's family moved from England to America in the 17th century.
In 1802 Nathaniel Bowditch published the <u>American Practical Navigator.</u>
That work remains the authority on Navigation in America to this day.

*Nathaniel Bowditch defines **KNOT** as follows: Noun,*
A unit of speed equal to 1 nautical mile per hour.

I know I am being repetitious.

Relax and let it sink in.

On charts, minutes of latitude are calibrated on the left and right boundaries of the chart; 60-minutes per degree
1-nautical mile is equal to 1-minute of latitude

Two Boats School – Two Circles of Time

1, 2, 3, & A, B, C

Daily Test
ANSWERS

Navigation Triangle

A = _____*Latitude*_____ on Charts

B = _____*Longitude*_____ on Charts

C = _____*Course Line*_____ on Charts

*Line of Sight Observation* in Celestial Navigation

The shortest distance between any 2-Points is a _____*Straight Line*_____ .

"We steer in a straight line to save time and fuel."

The shortest distance between any 3-points is a _____*Triangle*_____ .

"With 2-known factors in a navigation triangle we can find much more than just a third." "Remember: Common Sense saves time & lives."

Lesson #1 - Upper Grades Re-Test

Student's Name: _____

Date: _____

Grade in School: _____

Re-Test-Grade: _____

How many degrees are there in the following ?

#1: North = _____ .

#2: East = _____ .

#3: South = _____ .

#4: West = _____ .

#5: North-East = _____ .   { [000-degrees + 90-degrees] / 2 } = 45-degrees

#6: South-East = _____ .   { [180-degrees + 90-degrees] / 2 } = 135-degrees

#7: South-West = _____   { [180-degrees + 270-degrees] / 2 } = 225-degrees

#8: North-West = _____ .   { [360-degrees + 270-degrees] / 2 } = 315-degrees

#9: Write out Pythagorean Theorem . _____ .

#10: How many degrees are there in a Right Triangle ? _____ .

#11: Every Right Triangle has a certain kind of angle in it.
    That particular angle is always a _____ .

#12: There are how many degrees in a circle ? _____ .

#13: 1-Nautical Mile is = to _____ of Latitude ?

#14: To take a celestial observation.
    You, measure the celestial objects _____ above the _____ .

#15: Numerically Latitudes run _____ & _____ .

#16: Numerically Longitudes run _____ & _____ .

#17: There are how many degrees of Lat. ? _____-North & _____-South.

#18: There are how many degrees of Long ? _____-East & _____-West.

#19: Vertical lines on a chart stretching N. & S. {like this ⬤} are: _____

#20: Horizontal lines on a chart stretching E. & W. {like this ⬤} are:_____

Lesson #1 - Upper Grades Re-Test Answers

Student's Name: _____

Date: _____

Grade in School: _____

Re- Test-Grade: _____

How many degrees are there in the following ?

#1: North = *360-degees or 000-degrees*

#2: East   = *90-degrees*

#3: South = *180-degrees*

#4: West   = *270-degrees*

#5: North-East = *45-degrees*   { [000-degrees + 90-degrees] / 2 } = 45-degrees

#6: South-East = *135-degrees*   { [180-degrees + 90-degrees] / 2 } = 135-degrees

#7: South-West = *225-degrees*   { [180-degrees + 270-degrees] / 2 } = 225-degrees

#8: North-West = *315-degrees*   { [360-degrees + 270-degrees] / 2 } = 315-degrees

#9: Write out Pythagorean Theorem. $\{ A^2 + B^2 = C^2 \}$

#10: How many degrees are there in a Plane Right Triangle ? *180-degrees*

#11: Every Plane Right Triangle has a certain kind of angle in it.
     That particular angle is always a *{90-degree angle}* .

#12: There are how many degrees in a circle ? *{360-degrees}*

#13: 1-Nautical Mile is = to *{I-Minute of }* Latitude.

#14: To take a celestial observation.   *Angle*
     You, measure the celestial objects - *Height* - above the - *Horizon* -

#15: Numerically Latitudes run   - *North* - & - *South* - .

#16: Numerically Longitudes run -*East* - & - *West* - .

#17: There are how many degrees of Lat. ? *{90-degrees South}* & *{90-degrees North}*

#18: There are how many degrees of Long ? *{180-degrees East}* & *{180-degrees West}*

#19: Vertical lines on a chart stretching N. & S. {like this ⬤ .} are: *Longitudes* .

#20: Horizontal lines on a chart stretching E. & W. {like this ⬤ } are: *Latitudes.*

Lesson #1
Turtle Town Marina, Two Boats School

Test

Vocabulary

What are the meanings for the following abbreviations?

#1: L1 _____.

#2: L2 _____.

#3: D'Lat _____.

#4: D'Long _____

_____.

#5: M'Lat _____.

#6: NMs _____.

#7: 1-Knot equals how many feet per minute? _____.

Argument within Abbreviations

Let us examine my usage of abbreviations
in our last test.

Question #5: I used M'Lat for Mid-Latitude.

Captain Bowditch used "Lm" to represent
Mid-Latitude on page 581, Volume #2.
I will accept any answer on Q. #5 that you can argue a
reference to Bowditch with on the above.

Common Sense will prevail
over all situations where
language is an obstacle.

I certainly admire "Lm" representing Mid-Latitude.
My argument is. I was taught by my superior officers to
use any comfortable reference you want for Lm; M'L,
M'Lat, or Mid-Lat., etc. to represent Mid-Latitude.
Lm is found by adding L1 and L2 then dividing by 2.
"Always add L1 & L2 as whole #s to the 5th decimal."

Conversation Over: SNW

Lesson #1
Turtle Town Marina, Two Boats School
Test {{{Answers}}}

Vocabulary

What are the meanings for the following abbreviations?

#1: L 1 _____ Departure Latitude _____ .

#2: L 2 _____ Arrival Latitude _____ .

#3: D'Lat _____ Difference in Latitudes _____ .

#4: D'Long _____ Difference in Longitudes _____ associated with L 1 and L 2 _____ .

#5: M'Lat _____ Mid-Latitude _____ .

#6: NMs _____ Nautical Miles _____ .

#7: 1-Knot equals how many feet per minute? _100-ft

Test Prep for Lessons #5 and #6

Advanced Navigation

Reference: The last sentence on our previous page #42 introduces the concept of Latitudes and/or Longitudes as whole numbers. This is a new concept and is explained below.

"Always reduce latitudes and longitudes to whole #s to the 5th decimal when using trigonometry via Captain Soh Cah Toa."

This is a simplistic memory aid to advanced navigation.

Test Prep

Reduce Latitude **25' 50" N** to a whole number by dividing 60 into 50.

Latitude **25' 50"** reduced to a whole number for Captain Soh Cah Toa

$$50" = \frac{50}{60} = 60\overline{)50.00000}^{\,0.}$$

Test Prep Worksheet

25' 50"

Reduce 25' 50" N Lat. to a whole number for Capt. Soh Cah Toa. Drop a <u>ZERO</u> "0" every time you need a larger number to divide into.

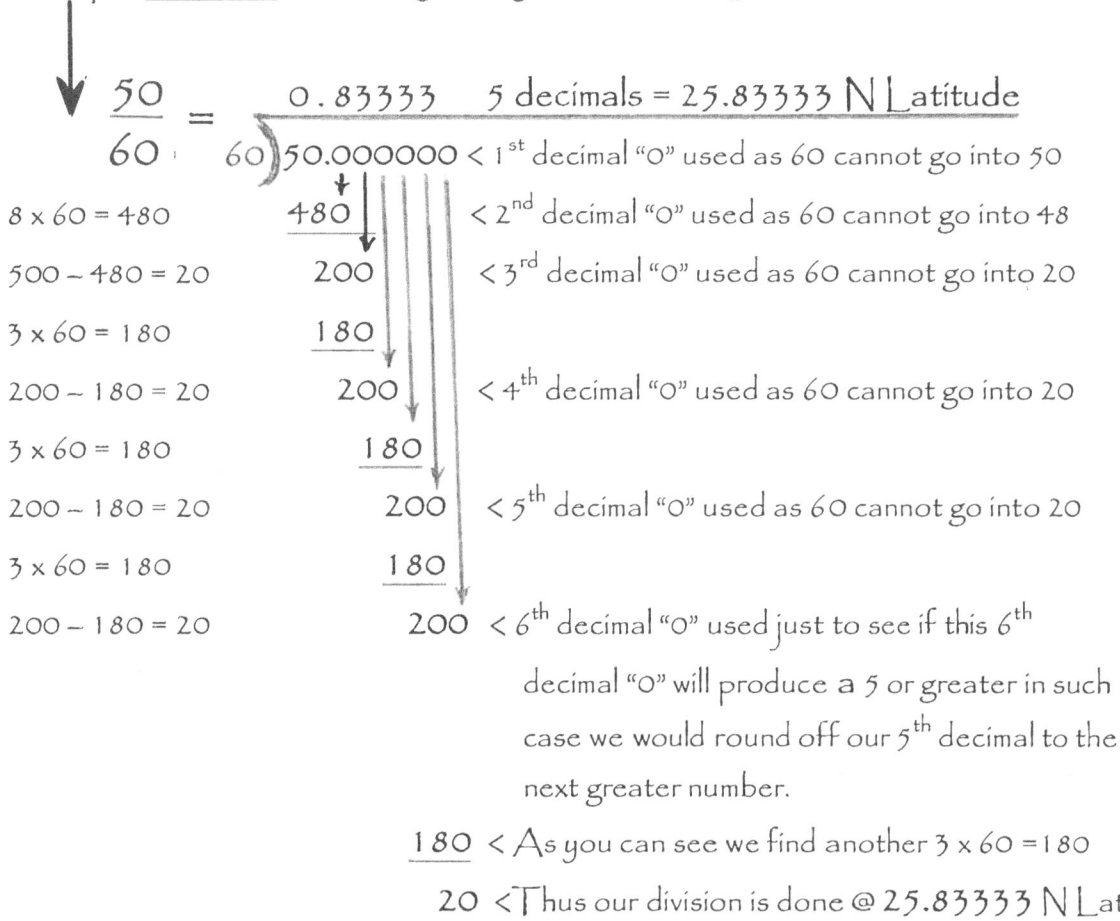

$$\frac{50}{60} = $$ 0.83333   5 decimals = 25.83333 N Latitude

60) 50.000000 < 1st decimal "0" used as 60 cannot go into 50

8 x 60 = 480   480 < 2nd decimal "0" used as 60 cannot go into 48

500 ~ 480 = 20   200 < 3rd decimal "0" used as 60 cannot go into 20

3 x 60 = 180   180

200 ~ 180 = 20   200 < 4th decimal "0" used as 60 cannot go into 20

3 x 60 = 180   180

200 ~ 180 = 20   200 < 5th decimal "0" used as 60 cannot go into 20

3 x 60 = 180   180

200 ~ 180 = 20   200 < 6th decimal "0" used just to see if this 6th

decimal "0" will produce a 5 or greater in such

case we would round off our 5th decimal to the

next greater number.

180 < As you can see we find another 3 x 60 = 180

20 < Thus our division is done @ 25.83333 N Lat

Answer = 25.83333' N Lat

Practice makes perfect and obviously anything very simple like adding and subtracting Lats and Longs as degrees and minutes you can sometimes do in your head.

Just remember you must reduce your Lats and Longs to whole numbers with 5-decmils when you use trigonometry via the

Captain Soh Cah Toa

in Lessons #5 and #6.

As elementary school students you first learn your numbers.
As middle school students you decipher your numbers.
As advanced students you prepare to earn a living.

For now familiarize yourselves with our abbreviations and nautical vocabulary associated with A, B, & C.

Study the change in vocabulary to Adjacent, Opposite, and Hypotenuse. These terms indicate that we are now entering trigonometry according to

Captain Soh Cah Toa.

The Navigation Triangle is the backbone of Advanced Navigation.

Great Circle Sailing

Now, I hope that you are not disappointed.
But, your next test involving a navigation triangle,
reducing latitudes and longitudes to whole numbers, or
other advanced navigation practices will be in lessons #5 & #6.

We use a Common Denominator of Nautical Miles
There are 60-Minutes in every 1-Degree of Latitude
1-Minute of Latitude = 1-Nautical Mile

If you want to be a Captain you have to know your Port and Starboard.

Left <<<< >>>>Right

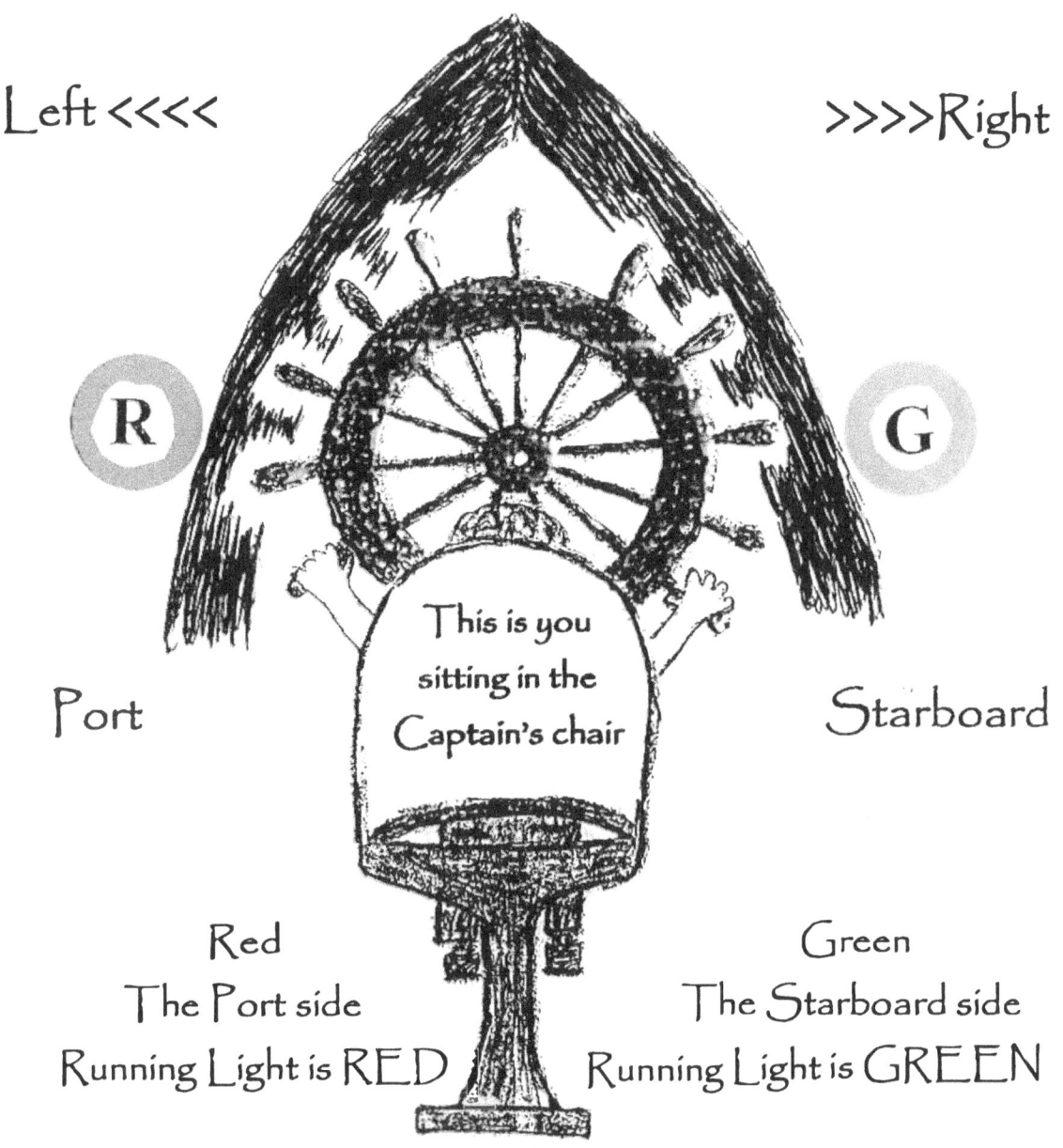

Port

This is you sitting in the Captain's chair

Starboard

Red
The Port side
Running Light is RED

Green
The Starboard side
Running Light is GREEN

I used to remember this by writing it down this way.

"Port, Left, and Red have fewer letters
than Starboard, Right, and Green."

"Do you know what I mean?"

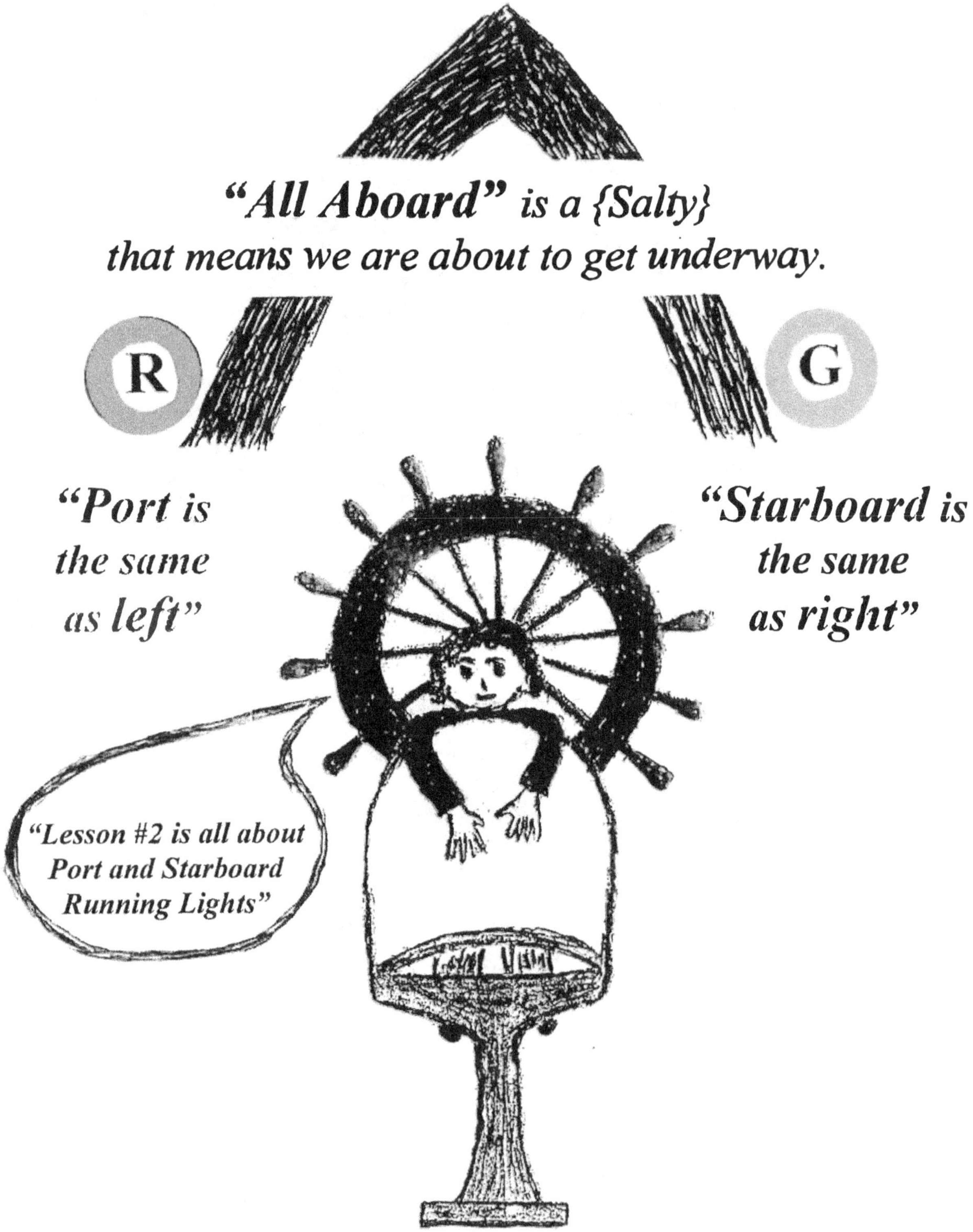

"All Aboard" is a {Salty}
that means we are about to get underway.

R   **G**

*"Port is
the same
as **left**"*

*"Starboard is
the same
as **right**"*

*"Lesson #2 is all about
Port and Starboard
Running Lights"*

Go ahead and write that salty down.
You can write. Right?

Repeat after me:

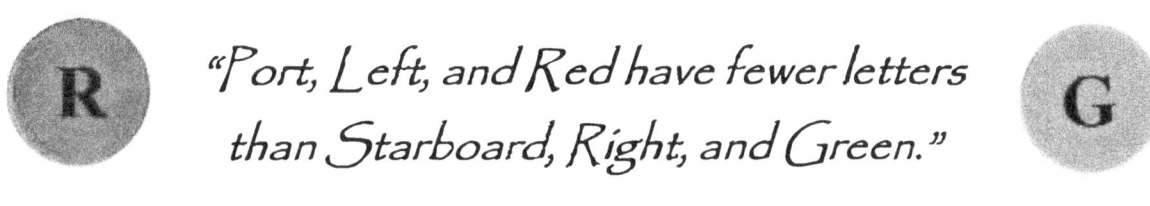

"Port, Left, and Red have fewer letters
than Starboard, Right, and Green."

Go ahead.

_____

_____

_____

_____

Did you have enough room?

Wait, wait, wait; don't go yet.

"Red on Right"

has "Right of Way"

You have already seen this Rule #15 on page 6.

<u>USCG Navigation Rule #15</u>: "When two power-driven vessels are crossing so as to involve risk of collision, the vessel which has the other to her own starboard side shall keep out of the way and shall, if the circumstances of the case admit, avoid crossing ahead of the other vessel."

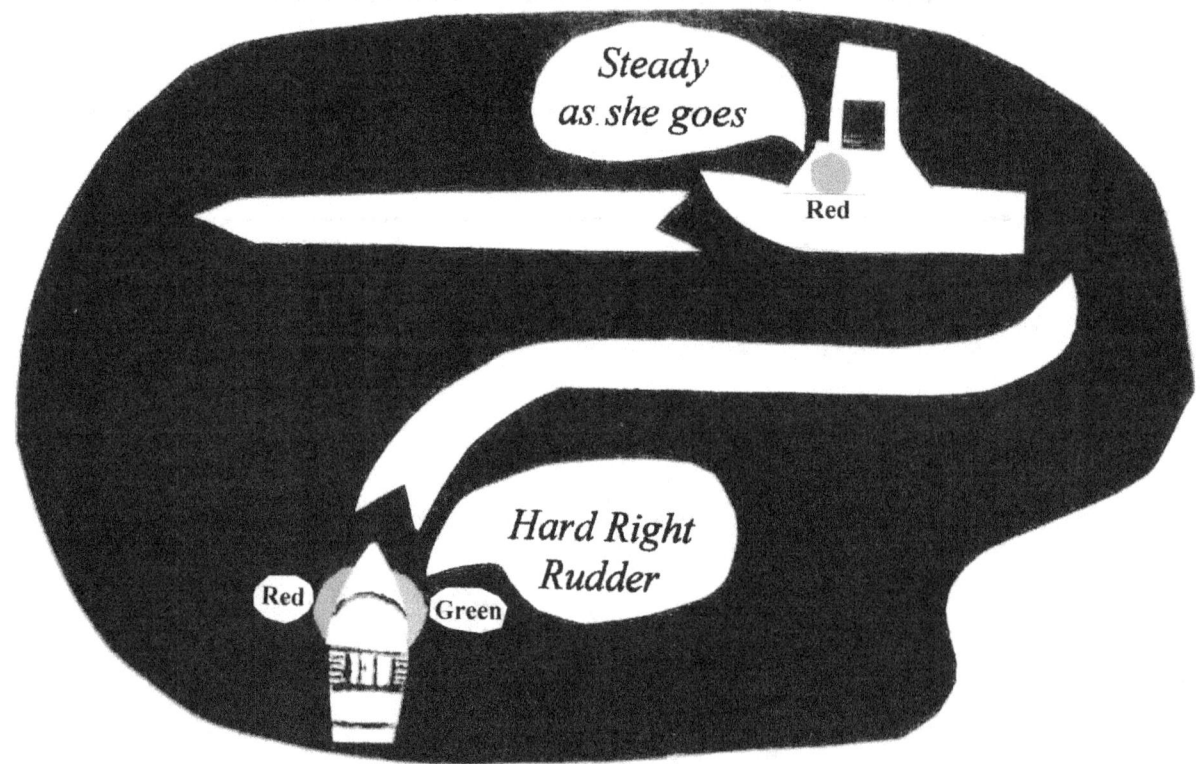

Wait, Wait, Wait; what are we going to do ?
We left during the day and I could see the land.
I could see the rocks. I could see the trees.
Now it is dark and all I see are lights.

Just remember Red, Right, Return!

When you leave port into open water you leave a
lighting system just like your boat.
You will leave with channel markers
Red on your Left and Green on your Right

Green   Red

G   R

Red   Green

Coming home you must remember;
Red, Right, Return.

The Channel Entrance
Blinks a Morse ode Letter "A": a {Dit-Da} in White
Dit = •   &   Da = —

Safe Water

Lesson #2

Port and Starboard

Grade in School: _____

Number of correct answers 1st time testing: _____

Number of correct answers 2nd time testing: _____

Number of correct answers 3rd time testing: _____

Check the box under the correct answer

G   **R**

#1: What color would this vessel's Running Light be? ☐ ☐

#2: What color would this vessel's Running Light be? ☐ ☐

#3: What color would this vessel's Running Lights be? ☐ ☐

#4: What color would this vessel's Running Light be? ☐ ☐

#5: What color would this vessel's Running Light be? ☐ ☐

Speed X Time = Distance

Just Remember

This is you sitting in the Captain's chair

Red — Green

Left << have fewer letters than >> Right

Port — Starboard

"See you in Lesson #2."

Lesson #2

Answers

Port and Starboard

Grade in School: _____

Number of correct answers 4th time testing: _____

Number of correct answers 5th time testing: _____

Number of correct answers 6th time testing: _____

Check the box under the correct answer

G   R

#1: What color would this vessel's Running Light be?   [✓]   [ ]

#2: What color would this vessel's Running Light be?   [ ]   [✓]

#3: What color would this vessel's Running Lights be?   [✓]   [✓]

#4: What color would this vessel's Running Light be?   [✓]   [ ]

#5: What color would this vessel's Running Light be?   [ ]   [✓]

Speed X Time = Distance

In the Sky
Red and Green

are the same as on our

Seas and Inland Waters

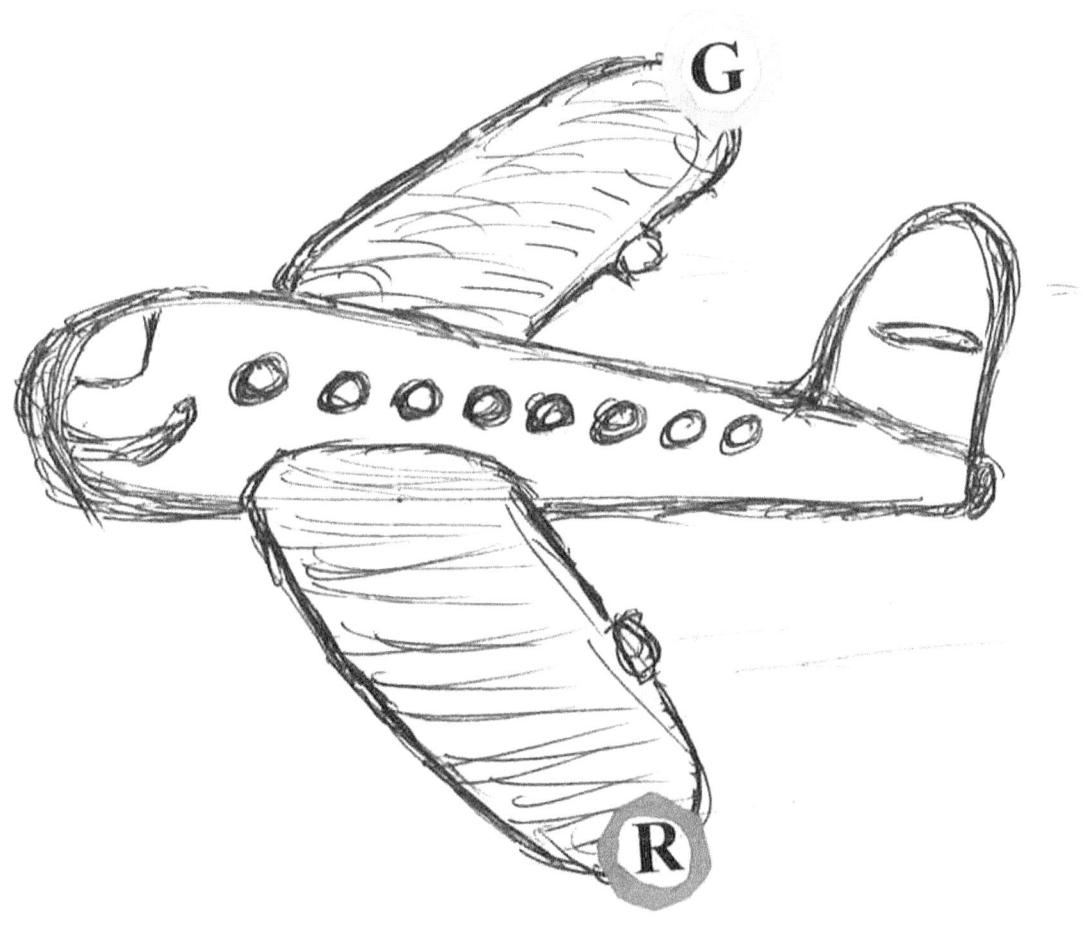

On the land
Red and Green
take a different stand

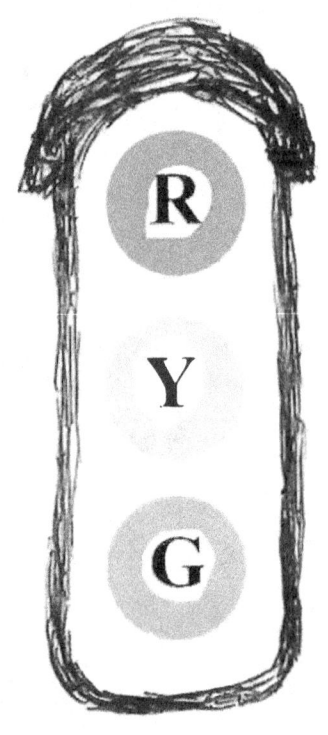

Red means STOP
Yellow means CAUTION light changing
Green means PRECEDE with Caution

{{{{{{ Roadway Traffic Light }}}}}}

Hi ! I'm your walking dividers.
I'll be walking you through
Lesson #3.

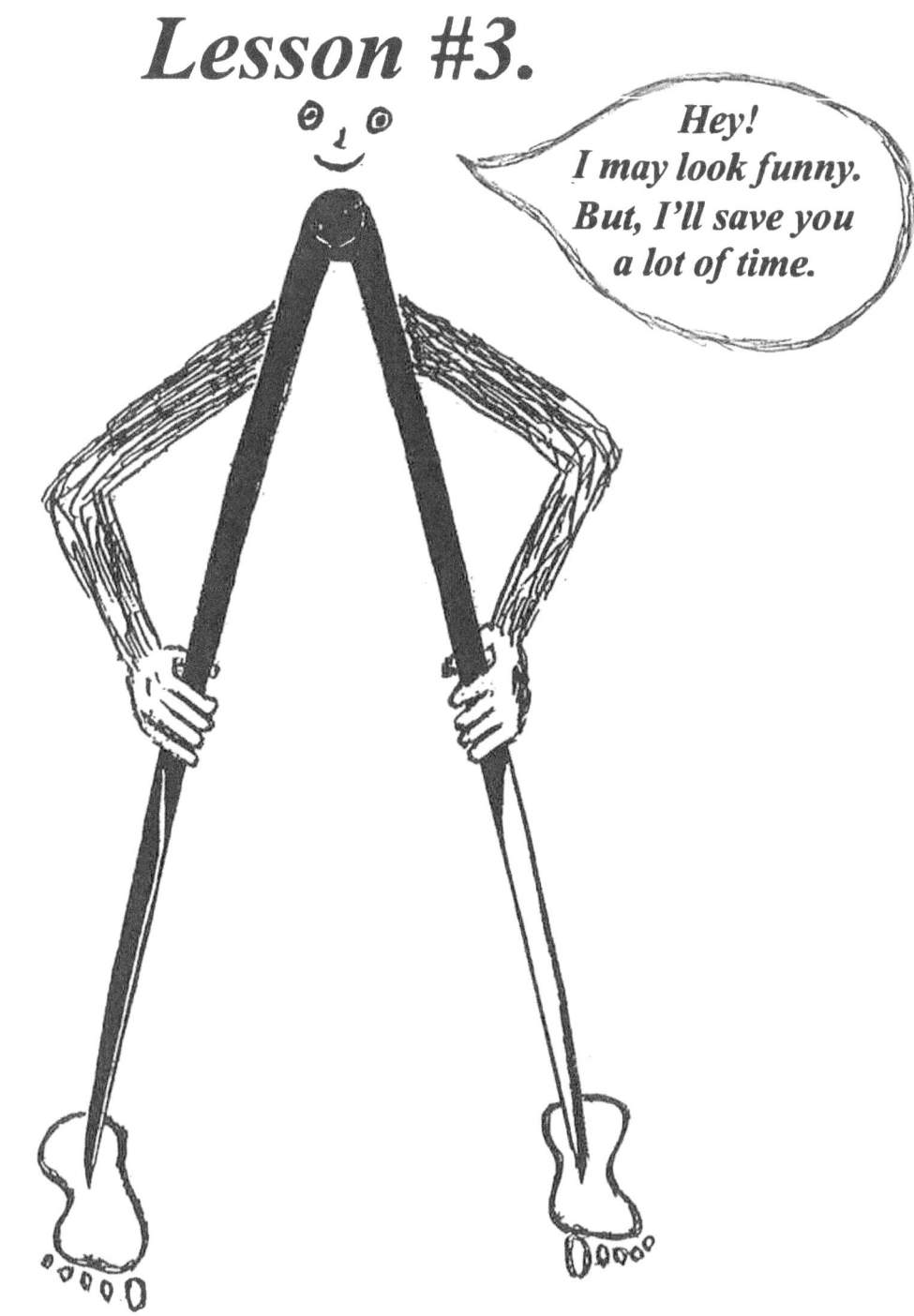

Actually; that is why I'm here, to save you time.
See you in Lesson #3.

Great Circle Study

Finding our Great Circle Course, Distance, Vertex, and Course Changes enroute the TRADITIONAL Way with Bowditch

Warning: Great Circle Sailings work best in Mid-Lats on east-west routes over 600-NM.
Remember: Sin, Cos, and Tan only work with degrees and tenths of a degree.

#1: {{ Answer: Distance is in Degrees. Must x 60 = Nautical Miles}}

#1: Cos Distance = {Sin L1 x Sin L2} +{Cos L1 x Cos L2 x Cos D'Long}
Page 1304, American Practical Navigator, Volume #1, Bowditch

#2: Tan Course = $\dfrac{\text{Sin D'Long}}{[\,\{\text{Cos L1 x Tan L2}\} - \{\text{Sin L1 x Cos D'Long}\}\,]}$

Page 1304, American Practical Navigator, Volume #1, Bowditch   #2: {{Apply "Compass Notation" to your Answer}}

#3: To calculate the Latitude of the Vertex   **Cos Lv = Cos L1 x Sin C**
Page 1305, American Practical Navigator, Volume #1, Bowditch

#4: To Calculate the <u>difference in Longitude</u> to the Vertex
Page 1305, American Practical Navigator, Volume #1, Bowditch

> **1st:** Calculate D'Long from Initial Long to Vertex
> **Sin D'Longv = $\dfrac{\text{Cos Cs}}{\text{Sin Lv}}$**

> **2nd:** Calculate the Longitude of Vertex:

> **Initial Long Minus D'Long headed East**
> **Initial Long Plus D'Long headed West**

#5: To calculate the <u>Latitude Lx</u> with Vertex known.
Page 1306, American Practical Navigator, Volume #1, Bowditch

> **Tan Lx = Cos D'Longvx x Tan Lv**

#6: Navigation Triangle   #6: {{Apply "Compass Notation" to your Answer}}
                    {Also applicable on legs of voyages under 600-NM's}

Capt Soh Cah Toa >> Sin < = $\dfrac{\text{Opposite}}{\text{Hypotenuse}}$, Cos < = $\dfrac{\text{Adjacent}}{\text{Hypotenuse}}$, Tan < = $\dfrac{\text{Opposite}}{\text{Adjacent}}$

p = D'Long in Nautical Miles = Cos Mid-Lat x D'Long in Minutes

Pages 575 – 620, American Practical Navigator, Volume #2, Bowditch, Reference Observation of p = DLo x cos Lm, Page 581.
Table 38 explaining DLo in Nautical Miles, Examine pages 595 and 604 for formulas used in The Sailings, Great Circle Sailing
{ < } = Angle of Interest

Great Circle Study

Up-coming Vocabulary

Captain Soh Cah Toa: Adapted K.I.S.S explanation; Keep It Simple Sailor: *The legendary Captain Soh Cah Toa is actually code to the Navigation Triangle.*

Great Circle Sailing: *Before we actually start working with Great Circle formulas imagine holding the world in your hands; one hand on the North Pole and one hand on the South Pole. The general current encompassing the North Atlantic travels clockwise with the Gulf Stream heading north from Florida and the Trade Winds heading west from North Africa are the basic flow masters. There are many more currents to be aware of as they can increase or decrease your vessel's speed. Now notice earth's longitudes on the far north and far south get as close together as your fingers do in the palm of your hand. Captain Nathaniel Bowditch, 1773-1838, understood our earth as a mathematical sphere. Our classroom's Great Circle Study voyage is from the Bahamas to the Mediterranean. We will travel into the deep well to the north of our port to port Rhumb-line course. While we gain latitude the math within shorter distances between longitudes becomes formulated trigonometry. This is our Great Circle Study on mid-latitude east-west voyages over 600-miles. Our Great Circle distance actually adds up to be the shorter than a straight Rhumb-line port to port voyage, about 76-miles shorter. We will test said distances as we study Great Circle formulas. SNW*

Latitude: Bowditch, Vol. #1, Pg. 61, *Latitude, {L, Lat} is angular distance from the equator, measured northward or southward along a meridian from 0-degrees at the equator to 90-degrees at the poles. It is designated north {N} or south {S} to indicate the direction of measurement.*

Latitude: *Latitudes are invisible in nature. Latitudes are an excellent mapping system on charts starting at Zero-Degrees at the Equator to 90-Degrees North and to 90-Degrees South. Latitudes are combined with Longitudes to establish a nautical position on charts or GPS; Global Positioning System, doing so at a 90-degree angle. Every latitude travels east/west around the world differentiated only by longitudes that travel north/south in system and nature.* <u>*Elementary explanation by SNW*</u>

D'Lat: Difference in Latitude, *D'Latitude is the difference between L1 and L2 described in total minutes of latitude or D'Lat in NM's; nautical miles.*

L1: *The latitude at the point of departure.* L1 = 25*50'N or 25.83333*N

Examples of L1 and L2

L2: *The latitude at the point of arrival.* L2 = 35*56'N or 35.93333*N

Longitude: Bowditch, Vol. #1, Pg. 62, *Longitude is the arc of a parallel or the angle at the pole between the prime meridian and the meridian of a point on the earth, measured eastward or westward from the prime meridian through 180-degrees. It is designated east {E} or west {W} to indicate the direction of measurement.*

Longitude: *Longitudes are invisible in nature. Longitudes are an excellent mapping system on charts starting at Zero-Degrees at the Greenwich Meridian or Zulu Time to 180-Degrees East and to 180-Degrees West at the International Date Line, Fiji. Longitudes are combined with latitudes to establish a nautical position on charts or GPS; Global Positioning Systems. Every longitude travels north/south around the world differentiated only by latitudes that travel east/west in system and nature.* Elementary Expl.: SNW

D'Long: *Difference in longitudes associated with L1 and L2 described in full degrees and tenths of a degree; such that the difference in longitudes will be compatible with functions sine, cosine, and tangent.*

Navigation Triangle according to Captain Soh Cah Toa

The navigation triangle consists of an opposite side of our angle of interest, the adjacent side connects to the 90-Degree Angle where our Latitude and Longitude meet, and the hypotenuse is our third side explained in Nautical Miles {our course}.

In the Navigation Triangle
we use a Common Denominator of Nautical Miles where
1-Minute Latitude = 1-Nautical Mile
The formula for these parts of D'Long is:

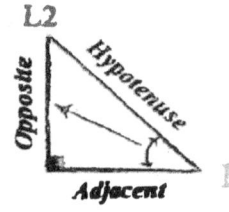

p = D'Long in NM's = Cos Mid-Lat x D'Long in Minutes = Nautical Miles
Page #581: American Practical Navigator, Volume #2

Mid-Latitude: *lm, Lm, or Mid-Lat is found by adding L1 and L2 as whole numbers to the 5ᵗʰ decimal then dividing by two. To transform eitherL1 or L2 to a whole number simply divide the minutes in question by 60.*

Rhumb-line: Bowditch, Vol. #1, Pg. #63, *Distance, as customarily used by the navigator, refers to the length of the rhumb line connecting two places.*

Vertex: Bowditch, Vol. #1, Pg. 271, In Great Circle Sailing: *The point of greatest latitude is called the vertex. For each great circle there is one of these in each hemisphere, 180-degrees apart.*

Manual Radar Collision Avoidance
Range and Bearing – Radar Plotting

Radar Range: 12-Nautical Miles
Rings: 2-Nautical Miles

Own Vessel Course: 360
360-Degrees = 000-Dgrees
Own Vessel Speed: 10-Knots

Your vessel remains at the center of the radar screen below.

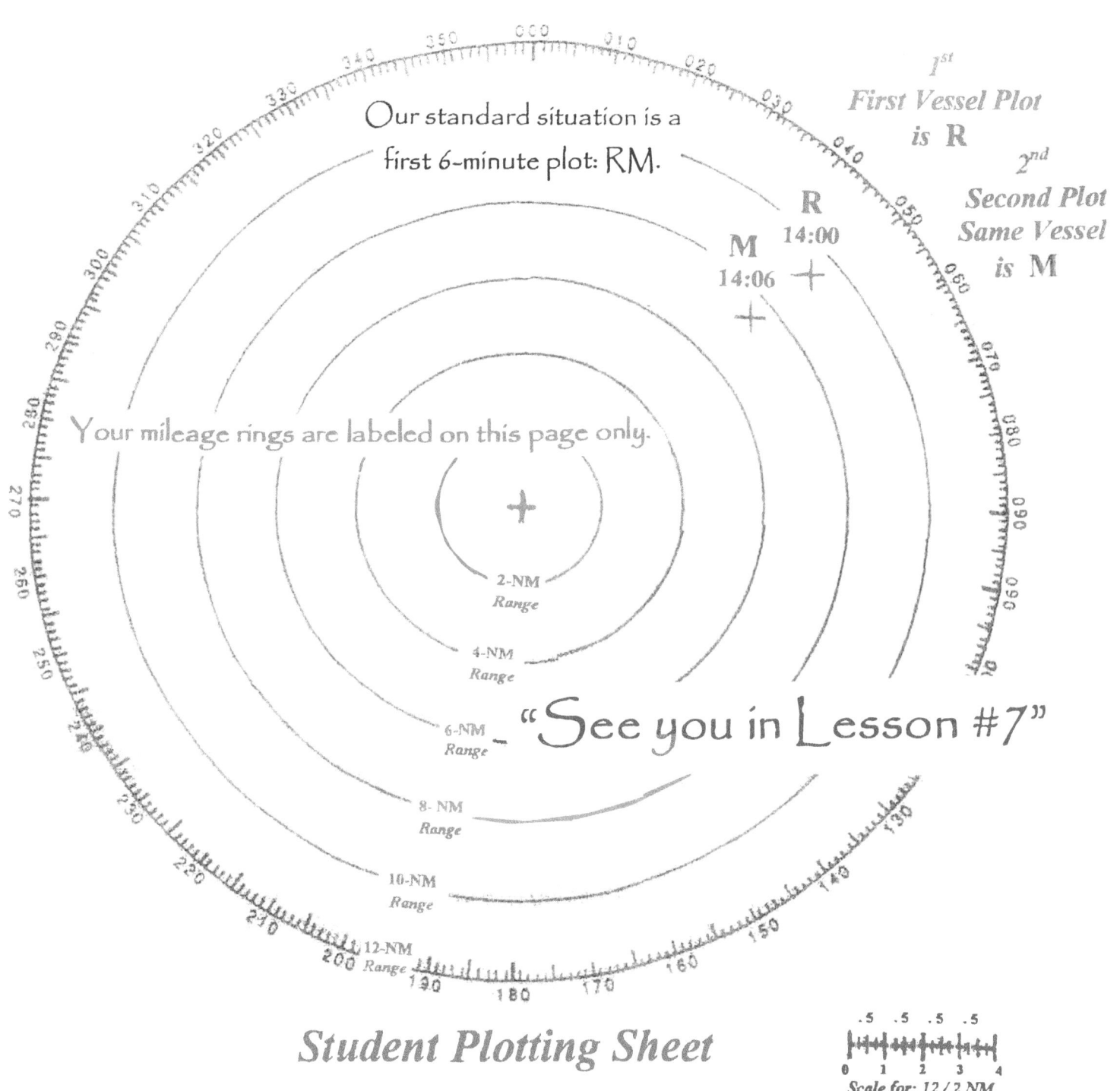

Our standard situation is a
first 6-minute plot: RM.

1st
First Vessel Plot
is R

2nd
Second Plot
Same Vessel
is M

R
14:00

M
14:06

Your mileage rings are labeled on this page only.

2-NM
Range

4-NM
Range

6-NM
Range

"See you in Lesson #7"

8-NM
Range

10-NM
Range

12-NM
Range

Student Plotting Sheet

.5   .5   .5   .5

0   1   2   3   4

Scale for: 12 / 2 NM

International Morse Code

uses a

Binary Alphabet
with only
2-Letters

Dit {.} & Da {-}

"See you in Lesson #8"

A = .-	Alfa
B = -...	Bravo
C = -.-.	Charley
D = -..	Delta
E = .	Echo
F = ..-.	Foxtrot
G = --.	Gulf
H = ...	Hotel
I = ..	India
J = .---	Juliett
K = -.-	Kilo
L = .-..	Lima
M = --	Mike
N = -.	November
O = ---	Oscar
P = .--.	Papa
Q = --.-	Quebec
R = .-.	Romeo
S = ...	Sierra
T = -	Tengo
U = ..-	Uniform
V = ...-	Victor
W = .--	Whiskey
X = -..-	X-ray
Y = -.--	Yankee
Z = --..	Zulu

Welcome Aboard

"Hi 5"

H = ... i = .. 5 = ...

Hi, I'm Nick Webster or

I wrote this **Two Boats School**, Turtle Town Marina, Lesson #1, because few years ago I had two life-threatening illnesses. As I got weaker I would write down all I could remember from my days at Sea to inspire young people to study math; division to trigonometry inside the Great Circle.

Put "Captain" on your resume'. U. S. Armed Forces, fisherman, shrimp boat, dive boat, treasure hunter, crew-boat, tugboat, deliveries, cruise ships, super tankers, tramp steamers, what-ever; no matter what else I say. The most important issue upon a captain today is the **Safety of Life at Sea**.

I started as a dock-boy for the Cove Marina in Norwalk, CT. in the 60's. In the 70's I was a rig-man aboard many vessels in the Gulf o Mexico. With Capt. Marlin Murphy of Morehead City I scalloped off Hudson Canyon east of Cape May. I shrimped with Capt. Huthmatcher off Shem Creek docking at Junior Magwood's of Sullivan Island. In the 80's I swordfished the Grand Banks of Newfoundland before testing for my first USCG 100-Ton Near Coastal and Inland Waters Captain's License studying with Capt. Zook of Morehead City, N.C. My first captain's berth was aboard the Ex-Pelorus; 159-feet, 26-staterooms, once a British Lightship renamed Langower. In the early 90's I captained 120-foot crew-boats with Tidewater and Tidex in the Oil-Patches of our Gulf of Mexico, Nigeria, Gabon, and repairs in the Ivory Coast. 5-years later I tested for my 500-Ton Master Oceans in New Orleans. Oceans mean USCG testing for Celestial Navigation. I left boats to work aboard ships boarding as an Able-Bodied Seaman with Coastal Tankship USA out of Houston, Texas. I up-graded every 5-years, traveled around the world two and a half times to retire with a 1600-Ton Master Oceans and a 2nd Officer Unlimited Tonnage Oceans. My last berth was as 3rd Officer, 08-12 Navigation Watch, Medical Officer, and Safety Officer aboard the M/V Ascension. In my last 10-years at Sea; all my Upper Level officer positions were staffed by the American Maritime Officers Union stationed at the Star Center in Dania Beach, Florida. I left the Sea in 2007 and was awarded a Medical Discharge from the Military Sealift Command in 2009.

Respectfully yours in Christ @ Sea & @ Home.
With Peace of Mind.

Nick Webster

Credits:

I thank every teacher I have ever had.
I especially thank my first and second grade teachers;
Mrs. Warren and Mrs. Lawrence.

That was back in Topsfield, Massachusetts in the 1950's.

I remember asking myself, "Why?".

Credits:

#1:

Navigation Rules
{International – Inland}
U.S. Department of Transportation
United States Coast Guard
2100 Second Street
Washington, D.C. 20593-0001

#2:

The American Practical Navigator
Originally by Nathaniel Bowditch
1773-1838, Born Salem Mass.
National Imagery and Mapping Agency
Lighthouse Press, Annapolis, MD

#3:

Pub. No 229, Vol. 1
Sight Reduction Tables for Marine Navigation
Defense Mapping Agency
Hydrographic / Topographical Center
Washington, DC

#4:

The Nautical Almanac

Washington	London
US Naval Observatory	Her Majesty's
Secretary of Defense	Secretary of State for Defense

#5:

Pub. 143, Sailing Directions {Enroute}
Revised through: Notice to Mariners
Lighthouse Press
Annapolis, MD

#6:

United Kingdom Hydrographic Office
Admiralty Way, Taunton, Summerset
TA9 4EX United Kingdom

#7:

Dutton's Nautical Navigation, 15th Ed., Naval Institute Press

#8:

American Maritime Officer's Union & RTM STAR Center
2 West Dixie Highway, Dania Beach, Florida, 33408

#9:

Captain Joe Lobo's, Deck License Program
www.uscgexam.com

Credits

Our United States Armed Forces

United States Army
United States Navy
United States Marines
United States Air Force
United States Coast Guard
United States Merchant Marines

Thank You
NASA

Thank you "One and All" for your service.

Thank you:
RTM Star Center
American Maritime Officer's Union
2 West Dixie Highway, Dania Beach, Florida 33408
My last employer

Thank you
Family, Friends, and Neighbors.

Romans 8: 37-39

"No, in all these things we are more than conquerors through Him who loved us. For I am sure that neither death nor life, nor angles nor rulers, nor things present nor things to come, nor powers, nor height nor depth, nor anything else in all creation, will be able to separate us from the love of God in Christ Jesus our Lord."

North Carolina
First in Flight
1903

Team USA
Project: "Good Karma"
2018